RAND ARROYO CENTER

T0302665

Assessing Locally Focused Stability Operations

Jan Osburg, Christopher Paul, Lisa Saum-Manning,
Dan Madden, Leslie Adrienne Payne

Prepared for the United States Army

Approved for public release; distribution unlimited

For more information on this publication, visit www.rand.org/t/rr387.html

Library of Congress Control Number: 2014944648

ISBN 978-0-8330-8564-1

Published by the RAND Corporation, Santa Monica, Calif.

© Copyright 2014 RAND Corporation

RAND® is a registered trademark.

Limited Print and Electronic Distribution Rights

This document and trademark(s) contained herein are protected by law. This representation of RAND intellectual property is provided for noncommercial use only. Unauthorized posting of this publication online is prohibited. Permission is given to duplicate this document for personal use only, as long as it is unaltered and complete. Permission is required from RAND to reproduce, or reuse in another form, any of its research documents for commercial use. For information on reprint and linking permissions, please visit www.rand.org/pubs/permissions.html.

The RAND Corporation is a research organization that develops solutions to public policy challenges to help make communities throughout the world safer and more secure, healthier and more prosperous. RAND is nonprofit, nonpartisan, and committed to the public interest.

RAND's publications do not necessarily reflect the opinions of its research clients and sponsors.

Support RAND

Make a tax-deductible charitable contribution at
www.rand.org/giving/contribute

www.rand.org

Preface

Locally focused stability operations (LFSO), such as the Village Stability Operations (VSO) effort in Afghanistan, can increase stability in strategic locations by fostering security, sustainable development, and effective governance. However, there is no standard doctrinal format for assessing the progress and outcomes of these operations.

To improve the Army's ability to measure and assess LFSO, this report identifies, distills, and documents information from the literature and from interviews with more than 50 subject-matter experts. It outlines best practices for measuring and assessing the effects of LFSO, identifies critical pitfalls to avoid, and suggests useful approaches for developing an LFSO assessment framework.

Findings and recommendations from the report are especially relevant to the United States Army's Asymmetric Warfare Group, which is charged with supporting Army and joint force commanders by providing operational advisory assistance to enhance combat effectiveness and reduce vulnerabilities to asymmetric threats, as well as to the Army's Regionally Aligned Forces and Special Forces, which often operate at a point in the spectrum of conflict where LFSO can be an option. Because the findings and recommendations presented are applicable to a wide range of operational contexts, this information should also be of interest to other organizations within the Army, as well as to organizations within the Air Force, the Navy, the Marines, and the Intelligence Community that are charged with conducting similar assessments. Planners tasked with designing assessment frameworks, the operators who execute the designs, analysts who interpret the data, and commanders who rely on their results to make critical operational decisions all play a role in a successful assessment effort and can find value in the findings presented here.

This research was sponsored by the Army's Asymmetric Warfare Group and was conducted within RAND Arroyo Center's Force Development and Technology Program. RAND Arroyo Center, a division of the RAND Corporation, is a federally funded research and development center sponsored by the United States Army.

Contents

Figures

Tables

Summary

Locally focused stability operations (LFSO) are an important element of counterinsurgency and stability operations. These missions, exemplified by the ongoing Village Stability Operations (VSO) effort in Afghanistan, typically involve small teams of U.S. or host-nation (HN) forces that work toward fostering local security, sustainable development, and effective governance in strategically located villages or similar areas. However, there is no standard doctrine for assessing the progress and outcomes of LFSO. As a result, those in need of such tools have developed and employed a plethora of different assessment approaches and models, with varying degrees of success.

To address the lack of standard assessment doctrine, the United States Army's Asymmetric Warfare Group (AWG) enlisted support from RAND's Arroyo Center to help determine analytic frameworks, best practices, and metrics for measuring the effects of LFSO. Based on an extensive literature review and interviews with more than 50 subject-matter experts (SMEs), this report identifies, distills, and documents information that will help improve the Army's ability to measure and assess operations in which the impact of local atmospherics and other factors of influence make evaluation especially difficult. Although the study is focused on assessing operations at the local level, it also identifies critical pitfalls to avoid in the assessment process and presents useful methods for developing an assessment framework for multiple operational contexts.

Best Practices for Addressing Challenges Related to LSFO Assessment

The study team first developed a working definition of LFSO that reflected the desired means and outcomes (ends) of such operations: *LFSO are the missions, tasks, and activities that build security, governance, and development by, with, and through the directly affected community to increase stability at the local level.*

This definition allowed the team to identify and review historic examples, interview experts, and assess relevant literature to determine which outcomes and costs should be measured (metrics) and how measurements should be collected (methods). However, simply providing the Army with a new laundry list of metrics to compete with the thousands of other metrics proposed in the literature appeared to be less useful than distilling and discussing the underlying principles of the metrics and demonstrating how these principles might be applied to an LFSO mission in the context of a specific contingency.

The study identified a number of foundational challenges that commanders and assessment experts face when conducting LFSO assessments, along with recommendations for addressing those challenges. The challenges and recommendations are summarized in Table S.1.

The interviews and literature also yielded implementation-related guidance to consider when developing and using an LSFO assessment framework:

- **Assessments should be commander-centric.** If the assessment process does not directly support the commander's decisionmaking, it should be reexamined. A good assessment process should include an "assessment of the assessment" to determine whether it results in new command decisions or at least improves the quality of the existing decision cycle. If it does not, a redesign is called for.
- **Assessments should reflect a clear Theory of Change.** To adequately support the commander, the assessment team must not only understand his objectives, it must also understand the underlying Theory of Change—how and why the commander believes

Table S.1
Summary of LFSO Assessment Challenges and Recommendations

Foundational Challenges	Recommendations
Assessing the impact of stability operations in a complex environment is not easy.	• Identify the challenges, take a deep breath, and forge ahead.
Doctrine and training fail to adequately address the complexities of assessment and appropriate skill sets.	• Prioritize assessment-related doctrine/ training. • Institutionalize the assessor role. • Assign individuals with the "right" personality traits. • Elicit SMEs to fill in contextual gaps. • Allow CONUS analysts to deploy to gain operational grounding.
Stakeholders (the United States, coalitions, HNs, NGOs, etc.) may have competing visions of stability.	• Establish an interagency/international working group to identify a set of variables across all lines of effort (security, governance, and development). • Develop an off-the-shelf assessment capability that uses a standard framework and is accepted by the broader stakeholder community.
There is a strategic- vs. tactical-level challenge: too much aggregation obfuscates nuance; too little can overwhelm consumers.	• Present results in a way that efficiently and clearly summarizes but can support more detailed exploration of data should the need arise.
Assessments sometimes rely on invalid or untested assumptions about causes and effects.	• Avoid drawing hasty conclusions by identifying/ documenting and testing/ validating assumptions. • Adjust the Theory of Change accordingly.
Bias, conflicts of interest, and other external factors can create perverse incentives.	• Triangulate to validate, using observable indicators, devil's advocacy, ratios, and other multiple-sourced methods.
Redundant reporting requirements and knowledge-management challenges impede the assessment process.	• Ask for data less frequently but require more in-depth responses, or ask for data more often but use less onerous questions. • Provide direct benefits (e.g., tailored products) to those who process data to validate and motivate.
Continuity of the assessment process can be difficult across deployment cycles.	• Plan for personnel turnover (training, documentation, knowledge management).
Assessment planning often ignores HN perspectives.	• Invite HN participation in assessment planning and execution. • Carefully consider hidden agendas.

NOTE: CONUS = continental United States; NGOs = nongovernmental organizations.

the tasks that have been laid out will result in the desired end state. A clearly articulated Theory of Change allows the assessment team to identify the appropriate inputs, outputs, and outcomes to measure and also enables it to determine whether critical assumptions built into the concept of operations may, if proven faulty, require the commander to adjust the campaign plan. This clear Theory of Change can also help the assessment team identify the truly critical outcomes to measure, so that it needs to drill down into more granular data only when the expected relationship between outcomes and inputs does not hold.

- **Assessments should seek to triangulate the truth.** Assessment teams should fully exploit the data and methods available, leveraging the strengths of a given source against the weaknesses of others, to triangulate ground truth. This should not be mistaken for simply haphazardly collecting as much data as possible to understand a particular trend. Analysts should take the time to understand the advantages and disadvantages of methods, including case studies, structured interviews, and regression analysis. When the resources are available, commanders would be well served by building an assessment team that exploits the healthy tension between quantitatively and qualitatively focused analysts. Commanders should also ensure that the team is led or overseen by someone they see as a trusted agent who can ensure that their methodological perspectives are synchronized to support their objectives, rather than simply being a science project that feeds reporting requirements of higher headquarters.

A Nested Framework for Setting Assessment Priorities

A number of assessment frameworks have been used to measure success in recent military operations. While there are numerous pros and cons related to each type of framework, there is one foundational principle for framework developers to consider: *hierarchical nesting*. Figure S.1 illustrates this concept.

Figure S.1
A Nested Hierarchy of Assessment

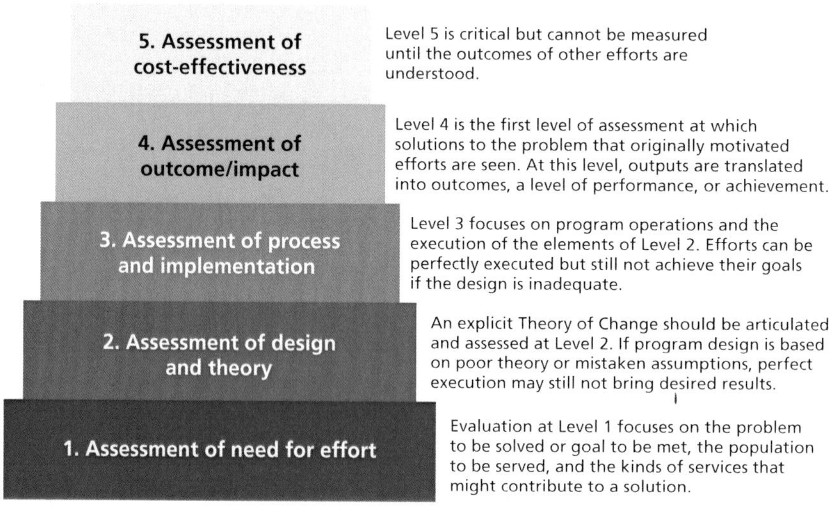

SOURCE: Adapted from Figure 7.1 in Paul et al., 2006, p. 110.
RAND *RR387-S.1*

In this hierarchical framework, the levels nest with each other, and solutions to problems observed at higher levels of assessment often lie at levels below. If the desired outcomes (Level 4) are achieved at the desired levels of cost-effectiveness (Level 5), lower levels of evaluation are irrelevant. However, when desired high-level outcomes are not achieved, information from the lower levels of assessment must be available to be examined.

A Recommended Course of Action for LFSO Assessment

To demonstrate the practical application of its assessment recommendations, the study team developed a comprehensive LFSO scenario for a notional African country. The team then designed an operational plan for stabilizing the local area and an associated assessment process consisting of the following steps:

1. Identify the challenges specific to the scenario.
2. Establish the Theory of Change behind the planned operation to help document the expected results and describe how activities and tasks are linked to those results.
3. Determine metrics and how/when to collect them.
4. Set up processes for data analysis (including aggregation) and communication of results.
5. Develop options for briefing leadership and stakeholders on the assessment plan.

Currently, there is no standard format for assessing progress in stability operations. This report provides a point of reference for anyone tasked with such an endeavor.

Acknowledgments

The authors would like to thank the leadership and staff of the Asymmetric Warfare Group for providing vital guidance and support for this project. We also would like to express our gratitude to the many subject-matter experts we interviewed for taking time out of their busy schedules to share their insights with us. We are especially indebted to the United States Army Africa leadership for hosting our visit to their facility, and we truly appreciate the outstanding practical support provided by the AWG liaison officer.

We would also like to thank our reviewers, Daniel Egel of RAND and Nathan White of the National Defense University, as well as Kate Giglio of RAND's Office of External Affairs. Their constructive contributions significantly improved the quality of this report.

Acronyms

AFRICOM	United States Africa Command
ALP	Afghan Local Police
ANSF	Afghan National Security Forces
AO	area of operations
AWG	Asymmetric Warfare Group
CAA	Center for Army Analysis
CM	Capability Milestone
COIN	counterinsurgency
CONUS	continental United States
CUAT	Commander's Unit Assessment Tool
DoD	Department of Defense
DSF	District Stabilization Framework
FM	Field Manual
HN	host nation
ICAF	Interagency Conflict Assessment Framework
INS	insurgent
ISAF	International Security Assistance Force
JP	Joint Publication
JTOSF-P	Joint Special Operations Task Force–Philippines
LFSO	locally focused stability operations

MOE	measure of effectiveness
MPICE	Measuring Progress in Conflict Environments
NATO	North Atlantic Treaty Organization
NGO	nongovernmental organization
OE	operating environment
ORSA	Operations Research/Systems Analysis
RFI	request for information
SIGACT	significant activity
SMART	Strategic Measurement, Assessment, and Reporting Tool
SME	subject-matter expert
TCAPF	Tactical Conflict Assessment and Planning Framework
USAID	United States Agency for International Development
USEMB	United States Embassy
VSO	Village Stability Operations

CHAPTER ONE

Introduction and Study Methods

Joint doctrine calls for the Army to execute Unified Land Operations in accordance with national policy. This policy generally focuses on the offensive and defensive capabilities required by combined arms maneuver. However, hybrid conflict and counterinsurgency (COIN) operations emphasize the importance of stability at the theater level and below. In this context, locally focused stability operations (LFSO), such as current Village Stability Operations (VSO) efforts in Afghanistan, can create stability through fostering security, sustainable development, and effective governance at the local level.

However, the success or failure of such efforts needs to be defined and measured. Currently, there is no standard doctrine for assessing progress in stability operations, though the *need* for assessments has been institutionalized in the operations process through doctrine (U.S. Joint Chiefs of Staff, 2011a). Further, many opine that existing guidance fails to adequately address how to design, plan, and execute such assessment (Bowers, 2013; Schroden, 2011; Zyck, 2011). While a variety of tools are available to assess the effects of LFSO, the complex nature of such operations makes assessment especially challenging. Theater-level events can influence progress at the tactical level, and while local context is key, theater-level assessments are ultimately required. In addition, local atmospherics and other indicators can be hard to quantify. Approaches vary even more when the ways in which the international community, nongovernmental organizations (NGOs), host nations (HNs), and other stakeholders measure progress in insecure operating environments are considered.

Recognizing the challenges of LFSO assessment and the shortcomings of current attempts, the United States Army Asymmetric Warfare Group (AWG) asked RAND's Arroyo Center to determine analytic frameworks, best practices, and metrics for measuring the effects of LFSO, including VSO in Afghanistan. This report documents the results of our efforts.

Locally Focused Stability Operations

It was necessary to clearly define LFSO at the onset of this project; specifically, LFSO needed to be differentiated from stability operations more widely, as well as from other kinds of operations. According to joint operations documentation, U.S. military doctrine defines stability operations quite generally as

> An overarching term encompassing various military missions, tasks, and activities conducted outside the United States in coordination with other instruments of national power to maintain or reestablish a safe and secure environment, provide essential governmental services, emergency infrastructure reconstruction, and humanitarian relief. (U.S. Joint Chiefs of Staff, 2011b)[1]

While there is no precise doctrinal definition for LFSO, the concept as we understand it involves small teams of U.S. or HN forces that embed in strategically located villages or similar locales to foster stability by generating and supervising locally based *security forces,* supporting sustainable *development,* and promoting effective *governance.* Our working definition is thus: *LFSO are the missions, tasks, and activities that build security, governance, and development by, with, and through the directly affected community to increase stability at the local level.*

Hence, LFSO are not just stability operations that reach down to the local level, they are stability operations that *leverage and enable local actors to create and maintain the building blocks for stability.*

---

[1] This definition is echoed in Army doctrine for stability operations (Headquarters, Department of the Army, 2008).

Project Scope

The obvious contemporary example of LFSO is VSO in Afghanistan. Historical parallels include U.S. Marine Corps efforts with Interim Security Critical Infrastructure in Helmand, the Civilian Irregular Defense Group experience in Vietnam, and the efforts of the British in Malaya. However, this study covered a generalized concept of LFSO, and its findings are intended to be applicable to a wide range of possible future LFSO in a multitude of possible contexts.

Methods and Approach

We asked several questions to help us begin to identify and distill information that will help improve the Army's ability to measure and assess LFSO:

- What are the characteristic elements of LFSO?
- What are the desired outcomes (ends) of such operations, and through what tools (means) can they be achieved?
- How can these outcomes and costs be measured (metrics), and how can these measurements be collected (methods)?
- How should these data be analyzed and the results communicated?

We conducted a literature review and interviews with subject-matter experts (SMEs) to answer these questions and inform the analysis leading to our proposed framework.

Our review of the literature involved a considerable array of both classified and unclassified sources,[2] including doctrine (both joint doctrine and Army doctrine), as well as nondoctrinal Department of Defense (DoD) handbooks, publications, and reports.[3] We also reviewed articles from relevant journals and periodicals, such as *Prism* and the *Small Wars Journal*, and reports and papers from other govern-

---

[2] This report is unclassified; classified materials were used only as sources for general principles or operational experiences apart from the classified details.

[3] Center for Army Lessons Learned, 2010.

ment agencies and organizations, including the Department of State, the United States Agency for International Development (USAID), and the National Defense University. We drew upon previous RAND research, as well as work from other research institutions and groups, including CNA Corporation, the Brookings Institution, the Center for Strategic and International Studies, the U.S. Institute for Peace, and the Military Operations Research Society.

Interviews with more than 50 SMEs[4] were begun concurrently with the literature review. The SMEs were identified in a number of ways: some were recommended by the sponsor (either specific individuals or organizations or units from which to seek a representative), some were individuals at RAND with relevant expertise, some were individuals contributing to the literature, and some were referrals from other SMEs and research-team members who identified representatives from organizations or units with recent relevant experience. The interviews were semistructured; they were conducted in person or by phone, and they lasted between 30 and 90 minutes.

Each interview was tailored to gather the most relevant information based on the SME's experience and area of expertise, but discussion often included the following questions: How would you define or bound LFSO? What historical examples should be included? What should be excluded? What assessment relevant to this area is being done, where, and using what methods or approaches? What lessons have you learned from managing/assessing local stability operations? How are indicators selected? What indicators are most difficult to measure? How often are assessment data collected? What is the proper balance between qualitative and quantitative assessment?

Once the literature review and the interviews were complete, notes were compiled and sorted based on recurring themes and key insights. During a series of team meetings involving all the authors of this report, key insights were distilled and synthesized. Input specific to assessment in the defense context and to assessment of LFSO

4 Interviewees were assured anonymity in order to foster candid discussion; their names and affiliations are therefore not cited in this report.

was integrated with preexisting team-member expertise on assessment, evaluation, and monitoring.

Organization of This Report

This report is designed with the practitioner in mind and is thus structured to facilitate ready access to answers of challenging questions. Chapter Two reviews selected assessment approaches and tools that have been used for LFSO assessment. Chapter Three notes the challenges associated with assessment and offers recommendations for practitioners to consider in moving forward. That chapter also recommends a framework for assessment design. Chapter Four presents a detailed example of the application of the framework in a notional West African LFSO scenario. Finally, Chapter Five presents our conclusions and summarizes our recommendations.

CHAPTER TWO

Review of Assessment Approaches in Stability Operations

As discussed in Chapter One, there is no current doctrinal assessment approach or tool for LFSO. In this chapter, we examine and review several approaches and tools that have been used in operations. First, we discuss three key dimensions of assessment and summarize insights on qualitative and quantitative assessment factors and standardization as presented in key literature sources. We also bring attention to the array of factors to be considered in LFSO assessment. We then examine a number of assessment tools that have been used in support of recent campaigns. We describe these tools and evaluate the way they may (or may not) be useful for assessing future LFSO. Finally, we present a short discussion of the role of survey research in such assessments.

Assessment Literature Review

Assessment approaches can be grouped along three dimensions: (1) tactical to strategic, (2) qualitative to quantitative, and (3) holistic to functional, as shown in Figure 2.1.

Tactical, Operational, and Strategic Assessments

In the tactical-to-strategic dimension, there are three levels at which assessment can be employed:

- **Tactical:** Assessment of a single village's stability, or the capability of an HN security forces unit.

Figure 2.1
Key Dimensions of Assessment

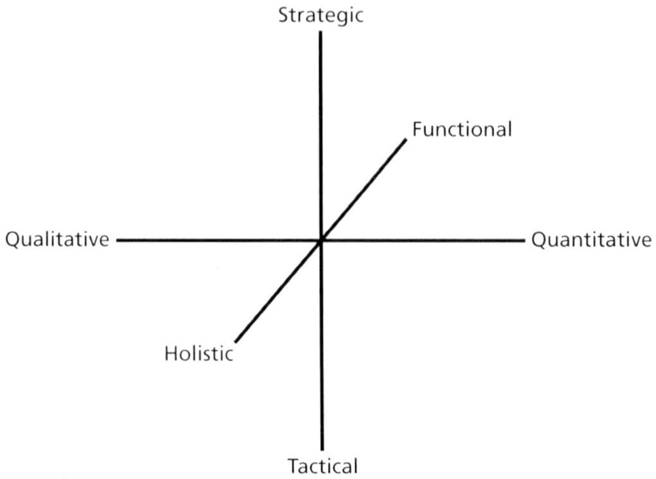

RAND *RR387-2.1*

- **Operational:** Assessment of whether a campaign plan is being successfully executed.
- **Strategic:** Assessment of the health of the U.S.-HN bilateral relationship, the behavior of regional actors, or whether a campaign is achieving its overall objectives.

In some cases, tactical assessments are simply aggregated upward to derive a campaign- or strategic-level assessment. In other cases, the operational and strategic levels require additional considerations beyond the summing up of tactical-level assessments, since the whole is often more than the sum of its parts, and tactical-level success does not always translate to operational- or strategic-level success. Complex operations often create situations in which it is possible to "win all the battles but still lose the war," and assessments across these levels must be mindful of this possibility.

Qualitative and Quantitative Assessments

For the qualitative-to-quantitative dimension, it is important to remember that quantitative approaches do not guarantee objectiveness, and qualitative approaches do not necessarily suffer from subjectivity. The judgments underlying a metric that is treated as quantitative can often be fundamentally subjective (e.g., the capability of an HN unit rated by its advisor along a quantitative scale), while some qualitative assessments can be perfectly objective (e.g., an advisor explaining that an HN unit is conducting independent operations). Another common issue is the treatment of observations that are communicated through quantitative measures as "empirical," while observations relayed through qualitative methods are treated as if they are somehow not empirical. Unfortunately, in practice, many of the quantitative metrics used in assessments are themselves anecdotal in that they reflect the observational bias of those reporting. For example, significant-activity (SIGACT) data used to report levels of violence in a particular area of a theater typically reflect only the reporting of coalition forces, sometimes including HN forces. The observations of HN citizens are omitted. Furthermore, some of the assessment approaches do not meet the criteria for being "quantitative" required by level-of-measurement theory (Stevens, 1946).

Currently, assessment literature is moving away from quantitatively focused methods and moving toward a mix of quantitative and qualitative considerations that more fully capture the dynamics at play in the complex COIN environment. Clancy and Crossett (2007) compare typical measures of effectiveness (e.g., counting the number of tanks destroyed and enemies killed in action) to methods more appropriate to the old game of Battleship—that is, methods that are antiquated and inadequate for interpreting and measuring the effects of operations on the modern battlefield. They attempt to broaden the considered field of assessment metrics, finding that insurgencies need to achieve several objectives to survive: sustainability, public legitimacy, and the ability to create chaos and instability. Based on historical and fictional scenarios, Clancy and Crossett conclude that military operations that counter these objectives seem to be highly successful and should be incorporated into the current set of measures of effectiveness (MOEs). Recent RAND research provides additional insight into the factors that play a role in COIN campaigns (Paul et al., 2013).

The Center for Army Lessons Learned (2010) describes several assessment tools commonly used to measure performance and effectiveness in stability operations. These tools measure both qualitative and quantitative factors, including (1) political, military, economic, social, information, infrastructure, physical environment, and time; (2) mission, enemy, terrain and weather, troops and support available/time available, and civil considerations; and (3) areas, structures, capabilities, organizations, people, and events.

The Center for Army Lessons Learned further describes a USAID-based method to integrate these assessments into operational and tactical planning called the Tactical Conflict Assessment and Planning Framework (TCAPF). The TCAPF is a questionnaire-based tool designed to assist commanders and their staffs in identifying and understanding the causes of instability, developing activities to diminish or mitigate them, and evaluating the effectiveness of the activities in fostering stability in a tactical-level (brigade, battalion, or company) area of operations (AO). The TCAPF contains a numeric scale that military staff can use to statistically measure local perceptions of the causes of instability and conflict in efforts to incorporate sociopolitical input in the military decisionmaking process.

Connable (2012) examines the methods used by theater-level commanders to assess ongoing U.S. COIN efforts. He finds that commanders largely rely on two primarily quantitative methods: Effects-Based Assessments and Pattern and Trend Analysis. These methods are highly centralized and tend to obfuscate many of the complexities inherent in COIN-centric conflict. As a result, military staffs resort to developing ad hoc assessment methods to capture insights from the tactical level but leave policymakers and the public confounded by how to interpret and assess ongoing efforts from a strategic point of view. Connable concludes that the decentralized, complex, and largely localized nature of COIN requires a much more nuanced approach to measuring progress (or failure). His report asserts that the most important recommendation for improving COIN assessment is to ensure that the overall process is transparent and includes contextual assessment. The process should incorporate explanatory narratives to fill in gaps and to compensate for centralized quantitative assessments that are often inadequate, inaccurate, and/or misleading.

Campbell, O'Hanlon, and Shapiro (2009) also suggest that metrics for measuring success in COIN environments are highly localized and situation-dependent. They further caution against using the same metrics to make judgments about different conflicts. The measurement focus in Iraq (trends in violence), for example, has not readily translated to Afghanistan, where gauging progress in building government capacity and economic development has taken center stage in the fight against a recalcitrant enemy. Campbell, O'Hanlon, and Shapiro find that in Afghanistan the most important metrics are those that gauge progress in security, economics, and politics. However, these state-building metrics are difficult to assess in insecure environments. Some potential social metrics to consider in Afghanistan include tracking trends in the daily life of typical citizens (e.g., How secure are they, and who do they credit for that security? How hopeful do they find their economic situation? Is the government capable of providing basic social services? Do they think their country's politics are giving them a voice?).

Kilcullen (2009) supports the consensus view against heavily quantitative approaches that rely on one-dimensional inputs such as body counts, enemy attack rates, and the amount of resources to determine effectiveness, since higher-order consequences can be dominant (see the section on Theory of Change in Chapter Three of this report). For example, killing one insurgent might produce 20 revenge-seeking others; a low number of enemy attacks may indicate uncontested enemy territory rather than government-controlled area; and measuring supplies and personnel figures tells a story about force-generation capabilities but not necessarily about the quality of the force generated.

Kilcullen contends that most of these narrowly focused indicators "tell us what we are doing, but not the effect we are having." He recommends shifting the focus of measurement from inputs to second- and third-order outcomes with respect to the behavior and perceptions of the population, HN government officials, security forces (military and police), and the enemy. He advocates adding context by examining various surrogate indicators that may reveal deeper trends in the security environment.

Integrated, Standardized Assessments

Successful COIN depends on understanding the unique drivers of conflict in the affected region—one of the reasons why many in the assessment community support a more tactical-level effort to measure effectiveness. However, in a departure from experts who advocate for decentralizing measurement techniques, Meharg (2009) contends that there is a need for far more coordination and centralization among assessment efforts. She points out that there are a myriad of stakeholders actively engaged in stability operations. This leads to a diffuse set of methods for measuring effectiveness. Military forces, humanitarian agencies, and civilian police, as well as the development sector, have different objectives based on differing views on what constitutes progress. As a result, activities that lead to progress in the short term from a military standpoint (e.g., rewarding intelligence sources with food) may be at odds with the aims and objectives of an aid organization working in the same area (e.g., preventing malnutrition for all), and both may impede the long-term goals of the HN government (e.g., developing a self-sufficient food-production capability).

Meharg concludes that what is needed is a way of thinking about measuring the whole rather than the parts, and she lobbies for a more integrated, holistic, and standardized approach to measuring effectiveness that includes metrics used by all sectors involved in efforts to generate long-term democratic peace and security. Such an approach might still draw upon the effects-based assessments commonly used by the military, but it would expand the method to other sectors to map a larger set of contributors to progress across multiple critical stability domains. Centralizing "effectiveness" data could inform intervention operations at many levels and provide baseline data to measure effectiveness in the longer term, something, Meharg contends, that the international community is not yet capable of.

In a similar vein, Banko et al. (undated) examine measurement practices used by North Atlantic Treaty Organization (NATO) members in current and recent theaters of operation to identify a potential set of core indicators that could provide a macro-level campaign assessment of specific interventions. With respect to concepts, approaches, and quality, the study finds a high level of diversity within and between nations and NGO assessments and concludes that there is a need for well-defined and consistent terminology and for core measures that are consistent across mission assessments to allow for comparisons between mission outcomes.

Banko et al. also conducted a comprehensive review of databases and datasets of public organizations, such as the World Bank and the World Health Organization, and identified a set of 33 variables, vetted by SMEs, that describe seven principal components relevant to COIN-focused environments: demographics, economy, education, health, infrastructure, law/security, and social constructs. The study provides a statistical demonstration of the ability of these indicators to show macro change during efforts to establish stability in conflict zones.

Assessment: More Art than Science?

LaRivee (2011) provides a useful review of some of the current philosophies and methods of assessment used by practitioners in the field. His report summarizes 20 short articles that describe the complexities and common pitfalls associated with developing sound assessment strategies in dynamic operating environments. Although presented as a guide for developing effective frameworks, the varying prescriptions presented suggest that assessment design is more an art than a science. LaRivee's synopsis centers on several key points:

- Philosophies to live by:
  - Ensure that the assessment process stays focused on measuring stated objectives.
  - Ensure that the assessment process is transparent and inclusive of a wide range of perspectives (including that of the HN when feasible).
  - Insist on data accessibility and objectivity in assessment-team efforts.

- Assessment design:
  - Standardize assessment frameworks enough to avoid confusion, yet keep them flexible enough to adapt to a fluid operating environment.
- Metrics management:
  - Select quality over quantity when incorporating metrics.
  - Characterize and discriminate between metrics and indicators to better understand their role in the assessment process.
  - Recall that metrics are not impervious to manipulation and subjectivity.

Review of Assessment Tools

In this section, we examine several assessment tools documented in the literature as of early 2013 that have been used in support of recent COIN campaigns. For each, we provide a brief description and then discuss their utility in assessing LFSO. Some of these tools and methods fit into multiple categories of the assessment dimensions outlined in Figure 2.1, since they employ a mixed-methods approach (e.g., assessment of VSO) or are scalable from tactical to strategic levels (e.g., Measuring Progress in Conflict Environments [MPICE]).

Functional and Quantitative Tools at the Tactical and Operational Level

Capability Milestone Rating System

From 2005 to 2010, U.S. military forces in Afghanistan used the Capability Milestone (CM) rating system as the primary system for measuring the development of Afghan National Security Forces (ANSF) capabilities against end-state goals. The quantitative CM system assessed Afghan army and police unit capabilities on a four-point scale in which a one indicated a unit was capable of operating independently and a four indicated the unit was established but not yet capable of conducting operations without some level of external support (making this a quasi-quantitative approach). Experts criticized the CM rating system for being too rigid, kinetic-focused, and narrowly quantitative and for overstating ANSF capabilities (Mausner, 2010). Furthermore, critics found that CM was not a reliable predictor of Afghan army unit effectiveness because it relied on assessments from commanders who may have had an incentive to spin data, since poor HN performance is a reflection on the commander's ability to achieve mission objectives (Office of the Special Inspector General for Afghanistan Reconstruction, 2010).

The CM was later replaced by the Commander's Unit Assessment Tool (CUAT), which sought to incorporate more qualitative instruments when measuring capabilities and stability. The lesser emphasis of the CM on qualitative analysis—the CM and the CUAT wholly focused on assessing the development of ANSF—in addition to its disproportionate focus of assessing HN security force development rather than village stability development, rendered it unsuitable for LFSO. For LFSO, an assessment framework that measures village-level developments in security, economics, and politics is most needed. Evaluations of the HN LFSO team's suitability will have to be conducted, but the primary emphasis should be on assessing stability developments within the village.

Commander's Unit Assessment Tool

The CUAT was developed in 2010 to replace the CM as the primary system for measuring the development of ANSF capabilities against end-state goals. It sought to incorporate additional metrics and called for narrative-based data collection. Since it was incorporated into assessments as a quantitative input, the CUAT can be considered quasi-quantitative.

The CUAT framework was the main assessment tool used in Afghanistan from 2010 onward. It is very familiar to practitioners of stability operations despite its shortcomings, and many may be tempted to leverage it in the future. However, choosing a tool such as the CUAT because of its comfort and familiarity may erode the effectiveness of the assessments. The chosen assessment framework should be capable of measuring many of the important issues the CUAT fails to address, such as loyalty of the HN security force, unit history, and HN corruption (Mausner, 2010). Critics of the CUAT have also said that it fails to address the sustainability of the HN security force, which is integral to assessing stability. The issues that are not measured by the CUAT are in fact integral to the success of LFSO. For example, the United States would have difficulty assessing the impact made by the HN LFSO team if it was unsure how reliable and dedicated the team was to the LFSO mission. Additionally, declaring the LFSO mission a success would first require an assurance of sustainability to ensure that the village would not revert to its former state. Most importantly, both the CUAT and the CM primarily focus on assessing the development of the HN security force, whereas an assessment framework that measures village-level developments in security, economics, and politics is needed for LFSO. Evaluations of the HN LFSO team's suitability will have to be conducted, but the primary emphasis must be on assessing stability developments within the village.

Holistic and Quantitative Tools at the Tactical and Operational Level

Village Stability Operations Assessment Process

VSO in Afghanistan focus on the integration of population security, security-force capacity, governance, and economic-development lines of effort at the local level. VSO involve the development of local security capacity through the Afghan Local Police (ALP) program, which draws recruits from the same community that the ALP are employed to protect.

The assessment of VSO at Combined Forces Special Operations Component Command–Afghanistan and later at Special Operations Joint Task Force–Afghanistan[1] was framed around the lines of effort described above. Two sources of data were employed: surveys filled out by VSO teams (typically O-3–level special operations units, e.g., a Special Forces Operational Detachment Alpha) and independent public opinion surveys conducted in districts where VSO was being conducted, focused on the same general topics. The VSO team-leader surveys included a qualitative section in which the respondents could

[1] The VSO assessment process was designed and supported by RAND analysts embedded at these headquarters, including several of the authors of this report.

provide a narrative description of how they assessed progress across the identified lines of effort within their area of responsibility.[2]

The VSO assessments were designed to assess the effectiveness of these operations in stabilizing targeted villages and districts. Results were used to develop estimates of conflict dynamics over time and changing population attitudes toward security forces and the Afghan government, among other topics.[3]

Since it was specifically designed for a form of LFSO, the VSO assessment approach deserves special attention. The chief advantages of the VSO assessment teams are their holistic approach to understanding stability and their effort at building on both quantitative and qualitative data from mutually independent sources (i.e., local population and VSO team leaders), enabling efforts to "triangulate" ground truth by comparing the insights and inferences from different perspectives. The VSO assessments have two chief limitations: (1) they do not fully integrate qualitative data and insights from other stakeholders into a unified assessment framework, and (2) they are designed not as campaign assessments per se, but rather as program and operational assessments. They were not directly linked to the International Security Assistance Force (ISAF) campaign plan in a more than thematic way; for example, the role of VSO sites in supporting ISAF-specified "key terrain districts" was not explicitly addressed.

The VSO assessment process can be treated as a point of departure for future efforts, but any assessment in a new region with a new mission will necessarily have to adapt the model to a new context. Additional efforts could be made to strengthen qualitative and multistakeholder contributions to the assessment process.

Holistic and Qualitative Tools at the Tactical and Operational Level

District Stabilization Framework Assessment Process

The USAID Office of Military Affairs created the District Stabilization Framework (DSF). The DSF is based on the premise that to increase stability in an area, practitioners must first understand what

---

[2] Unpublished RAND research by Walter Perry.

[3] Interview, February 6, 2013.

is causing instability in the operating environment. It is a population-centric framework that encourages practitioners and relevant agencies in a given area to establish a common situational awareness. The DSF has four primary steps: (1) identify sources of instability and gain an understanding of the environment; (2) analyze instability data and separate population wants and needs from the actual sources of instability; (3) start with indicators of change in sources of instability, then design activities/programs to effect change; (4) monitor and evaluate outcomes.

U.S. and coalition military and civilian personnel in Iraq, Afghanistan, and the Horn of Africa have employed the DSF. Practitioners in the field have described it as the "current framework of choice."[4] Its focus on causation and indicators of change may enhance its appeal to LFSO practitioners wishing to demonstrate the methodical process by which stability is finally effected in a village. Compared with other frameworks, and because it was designed to assess events at the district level, the DSF appears to be the most attractive option for LFSO. However, its implementation may be problematic for an HN LFSO team that does not recognize the value of prioritizing and collecting sociocultural and perception data from the population, since this is a cornerstone of the DSF. The framework was created under the assumption that U.S. or coalition actors would be the primary collectors of information. While it is not a requirement for the collectors to be career analysts or skilled sociologists, a shared understanding of the value of population-centric data can evoke a strong commitment from collectors. An HN LFSO team would need to have the same understanding to employ and extract meaning from the DSF.

Holistic and Quantitative Tools at the Strategic Level

Strategic Measurement, Assessment, and Reporting Tool

The Strategic Measurement, Assessment, and Reporting Tool (SMART) was specifically developed for the United States Africa Command (AFRICOM). It is a data manipulation, scoring, weighting, and calculation tool that presents information in an Excel spreadsheet and claims

---

[4] Interview, February 14, 2013.

to enable leadership to look at "big picture" performance throughout the African continent on a high-level objective. It is also described as being able to evaluate lower-level objectives at a country or regional level (Capshaw and Bassichis, undated).

Although it is able to look at lower-level objectives, SMART's strong suit is its ability to look at big-picture performance across the African continent. The tool is primarily quantitative, and conclusions about performance are reached by inferring meaning from tallied scores. Much of its acclaim is due to its modularity and its ability to capture a lot of detail. Despite these attributes, SMART may lack some essential capabilities needed to assess LFSO. Cross-referencing matrix data may provide insights in comparative analyses across many countries, but assessing LFSO requires thorough analysis of a single site or village. Therefore, a mixed-method approach that also employs qualitative analysis and the examination of causal factors may be preferred.

Measuring Progress in Conflict Environments Metrics Framework

The MPICE assessment framework was created at a workshop hosted by the United States Institute of Peace in which hundreds of academics, government officials, military personnel, and others participated. The framework has an exhaustive set of metrics for assessing stability operations that enable practitioners to track progress from the point of intervention (imposed stability) through stabilization (assisted stability) toward transition to a self-sustaining peace. It divides these metrics into the following categories:

- Political moderation and stable governance
- Safe and secure environment
- Rule of law
- Sustainable economy
- Social well-being.

MPICE focuses on identifying conflict drivers and incorporates both qualitative and quantitative measures, with slightly more emphasis on the latter. Similar to the DSF, it prioritizes the collection and analysis of social science data, which may be useful to LFSO practitioners. The framework includes about 800 generic, quantitative outcome metrics that measure institutional capacities and drivers of conflict in the five categories listed above; however, the framework is intended to be tailored down from the full suite of metrics to those most relevant to a particular conflict and adapted to the specific cultural context(s) of the conflict. The categories are related to the three *essential building blocks of stability* discussed further in Chapter Three: enhanced village security, improved political processes, and economic growth and maturation. This parity of objectives would render MPICE a suitable assessment framework for LFSO if it were not for the same drawback presented by the DSF—a quantitatively heavy framework with a plethora of metrics may strain the capacity of the HN LFSO team. However, an additional strength of the framework is its multisource approach, explicitly identifying the sorts of sources that will best capture trends (e.g., expert knowledge, survey/polling data, content analysis).

Interagency Conflict Assessment Framework

A U.S. government working group co-chaired by the Department of State's Office of the Coordinator for Stabilization and Reconstruction and USAID's Office of Conflict Management and Mitigation created the Interagency Conflict Assessment Framework (ICAF). It differs from other assessment protocols in that it draws on existing conflict assessment procedures like the TCAPF (which later became the DSF) and others. It organizes these assessments into a common framework that allows U.S. government departments and agencies to leverage and share the knowledge gained from their individual assessments and provides a common interagency perspective on individual countries or regions.

Like many of the assessment frameworks mentioned above, the ICAF recognizes the utility in examining causation. It leverages social science expertise and lays out a process by which an interagency team would identify societal and situational dynamics that are shown to increase or decrease the likelihood of violent conflict. The ICAF also provides a shared, interagency strategic snapshot of the conflict for future planning purposes (Office of the Coordinator for Reconstruction and Stabilization [c. 2008]). This snapshot would be only marginally useful for LFSO, which aim to measure stability at the local or village level and would be better served by a framework that could assess the sustainability of security, political, and economic development. The ICAF has been described as a high-level U.S. interagency assessment protocol that operates at the strategic level.[5] However, for LFSO, operational- and tactical-level frameworks that can be utilized and understood by the HN LFSO team are most needed.

Joint Special Operations Task Force–Philippines Assessment

For several years, the Center for Army Analysis (CAA) has been deploying analysts to various theaters, including Afghanistan, the Philippines, and the Horn of Africa, to provide embedded analytic support. A CAA analyst embedded with the Joint Special Operations Task Force–Philippines (JSOTF-P) was tasked by the commander with assessing how effective the command was at pursuing its foreign internal defense mission. Foreign internal defense involves working "through and with" partner nation forces, rather than unilaterally, to achieve stabilization objectives. The JSOTF-P commander observed to the CAA analyst, "I don't know how to know when I'm done."[6] The commander was seeking to move his command beyond a focus on developing HN tactical proficiency to a focus on operational-level capabilities. In particular, he was interested in developing three kinds of capabilities: operations and intelligence fusion cells, casualty evacuation, and operational planning. The two principal lines of effort were pressuring violent extremist-organization networks and enhancing friendly networks.

One of the challenges for conducting assessments that was identified early on was data management. JSOTF-P had several years of situation reports from commanders archived in various formats but no way to look back through the data systematically, no integrated database. The analyst noted, "It took me a week to review six months of data to answer one question."[7] CAA constructed a database, along with data

---

[5] Interview, February 15, 2013.

[6] Interview, February 7, 2013.

[7] Interview, February 7, 2013.

standards, reporting requirements, and a mechanism for conducting quality assessment of data entries. It required 1,200 hours and two months to structure one year of data from six years of situation reports and storyboards from JSOTF-P's subordinate units.

The JSOTF-P commander was surprised by the results of the assessment of engagement efforts. His guidance was that engagements should be focused on the operational level, but reporting data showed that tactical engagements (e.g., marksmanship training) constituted more than 99 percent of all engagements. The commander would not have been able to see that without the assessment, although he did sense that his guidance was not being met. What shocked him was the disparity between direction and execution. As an example, the task force was still providing marksmanship training to a key Philippine training center, rather than mentoring the center in developing its own initiatives.

To assess progress against the Operation Enduring Freedom–Philippines mission, the command has moved away from a heavy reliance on measures of activities (although some are still tracked) and has introduced a structured interview of commanders by the assessment analysts. In general, commanders appear to have been more receptive to interviews than to the more typical requests for information (RFIs) that come from higher headquarters—although it may be unrealistic to expect either to receive much attention in highly kinetic environments. RFIs would typically be delegated to an executive or operations officer and then reviewed by the commander for endorsement. The operations and intelligence briefs most commands have are too short-term in outlook to be a good basis for assessments. The interview process becomes an occasion for the commander to step back from his day-to-day tactical concerns and reflect on broader trends. The results of those interviews are then treated as hypotheses that the JSOTF-P analysts assess using both open-source and classified data. The assessment is treated more like writing a term paper than the traditional metrics-heavy assessment approaches seen in Iraq and Afghanistan.

Other indicators tracked include Foreign Military Sales, the existence of intelligence fusion cells, and engagement with HN SMEs. Working together with USAID, the Department of Justice, and other government agencies, the command has tried to identify other indicators of when its efforts were "bearing fruit." Examples of positive effects include a project becoming self-sustaining (e.g., local or government contributions provide for operating costs) and changes in public perceptions of the professionalism of the military. To support these assessment efforts, a quarterly public-opinion poll is administered in rural areas of the Philippines.[8]

Lessons for LFSO assessments include the value of structuring reporting formats so that they can easily be integrated into a database for time-series analysis and the potential value of structured interviews of commanders in the field. Having higher-echelon analysts interview commanders and staff could balance the objectivity of the analysts with the context-rich insight of commanders on the ground. In turn, this approach would have to be balanced against resource constraints, including movement across the battlefield and, more importantly, commanders' and staffs' time.

Holistic and Qualitative Tools at the Strategic Level

International Security Assistance Force Campaign Assessment

In 2011, ISAF transitioned away from the quantitative Quarterly Strategic Assessment Review to a more qualitative model developed by the Afghan Assessment Group at the behest of General John Allen, then commander of ISAF. Problems with the old assessment process included an excessive focus on context-poor metrics (e.g., Is it good or bad that SIGACTs are increasing during clearing operations?) and inappropriate treatment of data (e.g., averaging ordinal-level data). The new assessment process was split in two, with separate campaign and strategic assessments. The campaign assessment was focused on progress against ISAF's campaign plan, while the strategic assessment was focused on strategic objectives and regional plans. Subordinate functional commands (e.g., NATO Training Mission–Afghanistan, ISAF Joint Command) were responsible for providing narrative assessments, buttressed by quantitative data where appropriate, for the campaign

8 The analyst noted that opinion on relevant issues do not shift quickly enough to merit quarterly polling, and the JSOTF-P poll would likely shift to a semiannual basis.

assessment. The ISAF staff was responsible for the strategic assessment. This approach placed greater reliance on commanders' judgment and value on the context provided by narrative, rather than the abstract numeric coding of judgments or context-poor microlevel data (e.g., SIGACTs) (Connable, 2012).

ISAF's movement toward more-qualitative assessments appears to reflect broader trends in the assessment literature and is also reflected in assessment trends in the United States Pacific Command. However, it should be noted that this approach has not been validated through independent analysis. Although it has been well received by many who have participated in both legacy and reformed assessment and many of the critics of the traditional assessment approach, there is insufficient evidence to judge whether insights from the process have resulted in better decisions or a more accurate (precise and unbiased) depiction of ground truth. This is not to say that it is not a better approach than past efforts, simply that there appears to be insufficient evidence on which to judge.

Differentiating the requirements for tactical-, operational-, and strategic-level assessments appears to be useful for those at operational-level headquarters conducting LFSO as part of a broader campaign plan, even where LFSO is intended to be the decisive effort. The renewed focus on centering the assessment process on the commander's judgment is also promising and perhaps especially relevant where resources are unavailable for independent analysis, as may be the case in future LFSO efforts.

Understanding Survey Research

Understanding the population is widely seen as central to success in irregular warfare and at a strategic level in conventional conflict as well—as stated by Odierno, Amos, and McRaven: "Competition and conflict are about people" (Odierno, Amos, and McRaven, 2013). Surveys can therefore be a component of assessing population-centric efforts like LFSO. A survey is "a systematic method for gathering information from [a sample of] entities for the purpose of constructing quantitative descriptors of the attributes of the larger population of which the entities are members" (Groves et al., 2009). In other words, surveys are a method for describing a population based on a sample of the population. Familiar types of surveys include public-opinion polling[9] on political issues, election campaigns, and consumer confidence levels. In underdeveloped and conflict-affected areas, surveys typically include questions about health, education, economic conditions, and (less common among civilian-sponsored surveys) levels of violence.[10]

Frequently, survey data gathered in a war zone are viewed with suspicion, as is the analysis that relies on such data.[11] Thus, leaders and decisionmakers should approach survey findings with the same measured skepticism with which they treat other single-source reporting, particularly in conflict areas. Leaders should view survey findings within the broader context of other intelligence sources at their disposal, including unit reporting and key leader engagements. They should also expect this effort to contextualize survey findings from the staff presenting them with data.

The credibility of survey research is increased when findings are confirmed through independent observations, an approach also known as "triangulation."[12] When survey research, key leader engagements, intelligence sources, and unit reporting all point toward the same findings, commanders can have much greater confidence that they are basing decisions on sound data. Divergence of these sources on points critical to the commander's plans provides motivation for additional analysis or the establishment of new intelligence and information requirements (e.g., Commander's Critical Information Requests and Priority Intelligence Requirements).

---

[9] *Poll* is almost synonymous with *survey*. *Survey* is a broader term that could include, for example, a survey of soil samples across households, whereas *poll* is restricted to individuals' responses to questions. *Poll* and *survey* also differ by who conducts them: surveys are typically conducted by governments and academia, whereas polls are typically conducted by a private-sector entity (Groves et al., 2009).

[10] For a useful overview of surveys conducted in conflict areas, see Bruck et al., 2010.

[11] Unpublished RAND research by Ben Connable, Walter L. Perry, Abby Doll, Natasha Lander, and Dan Madden, 2012.

[12] In geometry and navigation, *triangulation* is the process of determining the location of a given point based on the direction to two known points.

While the proper conduct of surveys requires significant knowledge and skills, one need not have a degree in survey research to be an informed consumer of survey results. Decisionmakers should ask certain questions when presented with analyses based on survey research. The following questions are based on recommendations by the American Association for Public Opinion Research (Zukin, 2012), augmented by considerations specific to conflict environments:

- Who did the poll? Who paid for it?
- When was it done?
- Who was surveyed?
- How were respondents contacted?
- What is the sampling error?
- Are the data weighted? If so, why, and are they weighted appropriately?
- How are the questions (and responses) worded and ordered?
- What independent evidence is there that the polling results are valid?

Being an informed consumer of survey research also requires understanding the process of survey production, illustrated in Figure 2.2, and an awareness of the errors and biases that can be introduced at each step.

Conclusion

Dissatisfaction with existing approaches for assessing stability operations has spurred a renaissance in the literature, resulting in contending schools of thought that broadly fall into one of two camps: quantitative (variable-centric) or qualitative (commander-centric). One reason for the difficulty of systematizing the measurement process may be that the doctrinal characterization of "stability" is as broad as the tasks associated with achieving it: Army Field Manual (FM) 3-07, *Stability Operations*, defines stability operations as the "various military missions, tasks, and activities conducted outside the United States in coordina-

Figure 2.2
Survey-Production Life Cycle

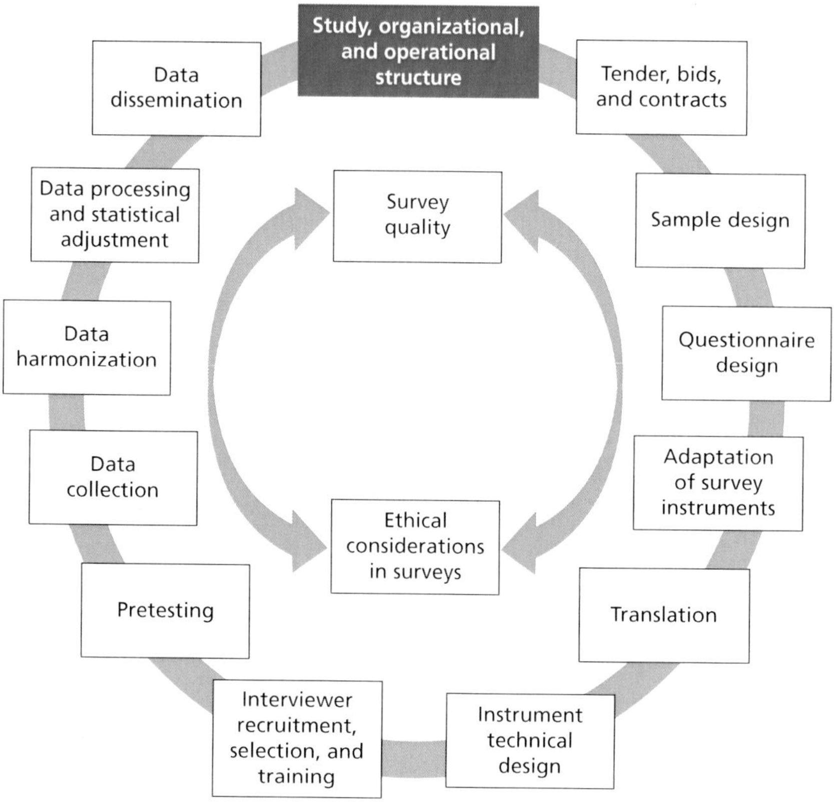

SOURCE: Survey Research Center, Institute for Social Research, 2011.
RAND *RR387-2.2*

tion with other instruments of national power to maintain or reestablish a safe, secure environment, provide essential government services, emergency infrastructure reconstruction, and humanitarian relief" (Headquarters, Department of the Army, 2008, p. viii). Approaches vary even more when considering the ways in which the international community, NGOs, HNs, and other stakeholders measure progress in contested operating environments.

CHAPTER THREE

Results: Insights and Recommendations

This chapter summarizes and synthesizes the key insights gained from our literature review and interviews. We highlight some foundational challenges associated with developing and conducting LFSO assessments and, where possible, offer recommendations based on the insights and best practices identified during the research. We also outline some implementation-focused guidance that commanders and their staffs should keep in mind when assessing LFSO. Although LFSO assessment is the focus of this research, some of the issues discussed can apply to other operational contexts as well.

Foundational Challenges

Lack of consensus on LFSO assessment principles and methods has led to an inability to measure LFSO outcomes with confidence and success. However, there are a number of common concerns regarding the foundations on which such assessment is based. These basic challenges must be addressed for the Army to create meaningful assessments:

1. The inherent complexity of LFSO missions
2. Limited assessment doctrine, training, and guidance
3. Competing visions of stability among stakeholders
4. The need to combine metrics and assessments across multiple areas and levels
5. Invalid or untested assumptions about causes and effects

6. Bias, conflicts of interest, and other external factors creating perverse incentives
7. Redundant reporting requirements and knowledge-management challenges
8. The difficulty of continuing the assessment process across deployment cycles
9. Failure to include HN perspectives.

While this is not a complete list, these challenges—many of which are generic to assessment of operations in conflict environments—are frequently cited as significant hurdles.

Challenge 1: Assessing the Impact of Stability Operations in a Complex Operating Environment Is Not Easy

> Realistically, no one, whatever background, deploys into the environment and has a perfectly complete understanding right away. We can't let the perfect be the enemy of the good.[1]

Findings

Nowhere in the literature reviewed or in our interviews was there a sense that assessment work is easy. Assessment processes and outcomes can fail under even the most permissive conditions. The fact that the U.S. military often has to conduct stability operations in complex security environments makes it all the more difficult to ensure the validity of metrics and methods used to evaluate measures of effectiveness. Concerns range from data collection (e.g., data are difficult to gather in a war zone),[2] to methods (e.g., certain methods of data aggregation are problematic),[3] to epistemological questions (e.g., whether objective measures can be meaningful outside a very narrowly defined time, place, and operational context).[4]

---

1 Interview, March 1, 2013.

2 Interview, November 17, 2012.

3 Interview, February 6, 2013.

4 Interview, January 29, 2013.

Recommendation

Complex inputs beget complex outcomes. That said, individuals charged with conducting assessments should not be discouraged. They should recognize the inherent challenges of the task at hand, take a deep breath, and forge ahead, with expectations managed accordingly.

Challenge 2: Doctrine and Training Fail to Adequately Address Assessment Complexities and Appropriate Skill Sets

> There is an enormous gap between how we are taught to conduct assessments and how we actually conduct assessments. (LaRivee, 2011)
>
> Computer error: please replace user.[5]

Findings

The consensus among most experts and practitioners is that those responsible for assessment often do not get sufficient higher-level guidance regarding how their tasks relate to the broader strategic mission of establishing stability in a country, and they get even less guidance on how to conduct assessments (Bowers, 2013).[6] Army FM 3-07 provides overarching doctrinal guidance for conducting stability operations and highlights the challenges of operating in a complex strategic environment. Yet little more than one page is devoted to conducting assessments in this context. What is more, FM 3-07 is geared toward "middle and senior leadership" (Headquarters, Department of the Army, 2008, p. iv) and provides little hands-on, practical guidance for tactical-level staff tasked with discerning what to make of an often fluid and dynamic area of responsibility. FM 5-0, *The Operations Process* (Headquarters, Department of the Army, 2010), digs deeper into conducting stability operations at the operational level and devotes an entire chapter to assessment, yet Schroden (2011) contends that its assessment guidance is rife with "deficiencies, contradictions and confusion." He notes, for example, that FM 5-0 argues both for and against

---

[5] Interview, March 1, 2013.

[6] Also interviews, January 9, 2013, and February 19, 2013.

detailed analysis, since two sections in the same chapter seem to stand in contradiction: Section 6-4 asserts that "committing valuable time and energy to developing excessive and time-consuming assessment schemes squanders resources better devoted to other operations process activities." Yet Section 6-41 acknowledges that "establishing cause and effect is sometimes difficult, but crucial to effective assessment. Commanders and staffs are well advised to devote the time, effort, and energy needed to properly uncover connections between causes and effects." FM 3-24, *Counterinsurgency*, provides some detail on useful measures and indicators for assessing progress in a COIN-centric environment, but it provides little guidance on how to design them, much less incorporate them into a comprehensive framework that enables an assessor to tell a meaningful story (Headquarters, Department of the Army, 2006, Chap. 5). Available resources may offer generalized guidelines for how assessments support operations, but they stop well short of fully preparing the practitioner for the messy conditions and shifting demands of real-time assessments.

Interviews with people in the training community suggest that predeployment training also does not focus enough attention on how to conduct assessments.[7] Rather, it typically focuses on basic soldiering skills (combat skills, drivers training, medical skills, vehicle maintenance, etc.) (162nd Infantry Brigade, 2011). Although COIN-centric campaigns in Iraq and Afghanistan have driven training requirements toward increasing language skills and cultural awareness, several interviewees criticized the training component at the Joint Readiness Training Center for failing to teach how to assess the complexities of the operating environment (OE). For example, the training concept for the Security Force Assistant Teams deploying to Afghanistan toward the latter part of Operation Enduring Freedom centered on cultural awareness, key leader engagements, language skills, and building rapport. Yet the seven-day program of instruction reproduced in Figure 3.1 shows that very little time is devoted to figuring out how to assess progress in these aspects of the mission.

---

[7] Interview, January 15, 2013.

Figure 3.1
Security Force Assistant Team Training Program of Instruction

Day 1	Day 2	Day 3	Day 4	Day 5	Day 6	Day 7
Language Skills (2 HR)	Language Skills (2 HR)	Language Skills (2 HR)	Language Skills (2 HR)	Language Skills (2 HR)	Language Skills (2 HR)	Language Skills (2 HR)
Welcome/ Security Force Assistance Overview (1 HR)	**Case study: Am Advisors Overseas (1 HR)**	**Case study: Multiplying by Zero (1 HR)**	**Case study: Canadian Army 14 Tenets (1 HR)**	**Case study: Afghanistan and Vietnam Template (1 HR)**	**Case study: Development Drives Instability (1 HR)**	Senior Leader Seminar and Police Advisor Skills
Course Overview (0.5 HR)	Islam Overview (3 HR)	**S-TT Aptitude Test (0.5 HR)**	Cross cultural commo/culture shock (1.5 HR)	Information Operations (1 HR)	Introduction to Key Leader Engagements (1 HR)	
Theater Framework (1 HR)	Afghanistan Country/Culture Overview (3 HR)	Government Overview (1 HR)	Rapport Building (1 HR)	VSO, ALP and APRP (1 HR)	FET/FST (1 HR)	
COMISAF COIN Guidance (2 HR)		ANA Overview (1 HR)	Influencing (1 HR)	Ethics/ Corruption (1 HR)	Media Awareness (2 HR)	
Mutual Perceptions/ Green-on-Blue (1 HR)		ANP Overview (1 HR)	Negotiations (1 HR)	HN Logistics (1 HR)	Vignettes (2 HR)	
Roles/Traits of Advisor (1 HR)		Interpreter Management (1 HR)	Development Skills (1 HR)	CUAT Overview (1 HR)		
			Training Foreign Forces (1 HR)			

SOURCE: 162nd Brigade, 2011.

Some assessment training is provided during Intermediate Level Education (the major's course), but assessments are typically actively conducted by lieutenants and captains. As one expert stated, "Unless a soldier's educational background happens to be in this field, then it is impractical to expect these guys to be able to conduct rigorous assessment."[8]

The lack of assessment-related training also extends to training the right people for the job and placing them in the right organization. While the limits of this study did not permit a rigorous determination of what to look for in the ideal candidate, it should be noted that assignments are often determined on an ad hoc basis, with staff officers, regardless of their skill sets, being placed in assessments billets. Even the talents of Operations Research/Systems Analysis (ORSA) specialists are often misdirected. While ORSAs generally have outstanding quantitative skills, they may have little background in qualitative methods (e.g., anthropology-based techniques) or in the regional context that is critical to understanding irregular warfare.

This general lack of guidance has both benefits and drawbacks. Minimal direction provides freedom of movement for staff with experience on the ground who may have in-depth understanding of how to develop metrics tailored to the realities they face in their AO.[9] However tactical-level staff may lack the assessment skills and expertise needed to accurately interpret data. To compensate, a command might request analytical support from SMEs in the continental United States (CONUS) who may not have the operational experience or situational awareness to put the numbers into context. This may result in flawed assessments that ultimately produce faulty conclusions about the state of the campaign.

In addition, one can only go so far with doctrine and training. Analysts must have the right kind of personality to do this type of work. Just as soldiers who possess certain characteristics, such as patience and emotional maturity, tend to be successful in conducting population-centric operations, assessment personnel may need to

---

[8] Interview, January 15, 2013.

[9] Interview, on February 7, 2013.

"win the hearts and minds" of the tactical-level operators on whom they rely for data (Military Operations Research Society, 2012). On the basis of our interviews and our experiences conducting assessments in Afghanistan, we found that soldiers, already tasked with an exhaustive amount of reporting requirements, are much more responsive to additional requests when the data collectors are sensitive to the demands of the soldiers' other daily tasks and priorities.[10]

Recommendations

We identified several options for overcoming these challenges. To address the doctrinal issue, one individual suggested that operational assessment "needs its own book."[11] He envisioned a baseline doctrine (a Joint Publication [JP] in the 5-0 series) that would then set the conditions for mission-dependent subordinate doctrinal coverage (i.e., covering specifics of COIN assessment in the COIN JP; assessment of stability operations in the Stability Operations JP, etc.). This doctrine should clearly articulate the skills required for conducting assessments to enable the training and education community to develop training objectives and standards. The Army should establish a proponent (e.g., the CAA) for the development of assessment capabilities (including doctrine, organization, training, materiel, leadership and education, personnel, facilities, and policy) to ensure that progress on this front in recent years is not lost and continues to mature.

It was also suggested that "assessment specialist" become a separate career field or skill identifier and be formally integrated within the professional military education system. Developing a professional assessor may be beyond the current limits and capabilities of the Military Occupational Specialty process.

Realistically, however, only so much training and education can be done in this regard. To address additional gaps, planners should

---

[10] Interviews, February 6, 2013, and February 22, 2013; also personal experience of Jan Osburg and Lisa Saum-Manning while embedded as assessment planners with the Combined Forces Special Operations Component Command–Afghanistan, Kabul, in 2010 and 2011, respectively.

[11] E-mail correspondence, June 25, 2013.

elicit the views of external SMEs to help give granularity to the context. Another option would be to have CONUS-based analysts do battlefield tours to observe and gain operational experience in order to better interpret data.[12] An indirect approach might be to bolster existing Red Team efforts and AWG-enabled fusion cells with additional billets to mitigate the "fog of war" and infuse additional rigor into the assessment process.[13]

Challenge 3: There Are Competing Visions of Stability Among Stakeholders (United States, Coalition, HNs, NGOs, etc.)

> USG [the U.S. government] is going to fail at this. Nobody owns it. The military wants to fight, AID [Agency for International Development] wants to do development, and DOS [Department of State] follows the ambassador. Nobody owns stability. To succeed, someone needs to own this.[14]

Findings

Often, a myriad of stakeholders are actively engaged in the OE. Military forces, humanitarian agencies, and civilian police, as well as the development sector, have different objectives based on differing views on what constitutes progress (Military Operations Research Society, 2012). Essentially, each stakeholder measures what is important to his own rice bowl rather than developing a common set of objectives for achieving stability.[15] One assessment expert commented, "I can't think of a single case where the whole interagency agrees about what the objectives are!"[16] As a result, activities that lead to progress in the short term from a military standpoint (e.g., providing fuel in exchange for information) may be at odds with the aims and objectives of an aid organization working in the same area (e.g., providing fuel aid for all),

---

[12] Interview, February 22, 2013.

[13] Comments of external reviewer, November 2013.

[14] Interview, February 15, 2013.

[15] Interview, February 14, 2013; also Downes-Martin, 2011.

[16] Interview, February 12, 2013.

and both may impede the long-term goals of the HN government (e.g., developing a self-sufficient fuel production and distribution market).

What is more, it is often difficult to measure the impact of U.S. military efforts in an OE that the actions of other stakeholders may influence in undetermined ways. One interviewee with recent experience conducting security-force assistance in an African country described a situation in which his platoon would occasionally come in contact with a hitherto unbeknownst Russian, Chinese, or Korean army unit that was conducting simultaneous training missions. It was never made clear whether there was a common end state or their efforts were diametrically in conflict. "It got to the point where we just had to say 'You stay over here east of this grid line so we don't shoot each other.'"[17] This type of dynamic not only makes it difficult to develop a common operating picture, it also muddies the water in terms of what activities are having an effect and how.

Recommendations

Some assessment-minded experts believe that what is needed is a way of thinking about measuring the whole rather than the parts and an assessment tool that is broadly applicable across contexts and can be accepted into the interagency fold (and ideally the international community) (Meharg, 2009). A more integrated and standardized approach to measuring effectiveness would include metrics used by all sectors involved in efforts to generate long-term democratic peace and security. This would require well-defined and consistent terminology and core measures that are consistent across mission assessments to allow for comparisons between mission outcomes (Banko et al., undated). Many also acknowledge that getting agencies to work together is difficult, even in the best of circumstances; in challenging operating environments, with little data and many conflicting objectives and opinions, it is especially difficult. For example, although NGOs can be a key resource, conflicting objectives and requirements (e.g., perceived neutrality) and historical tensions between the development community and the military can stifle coordination and collaboration, making it

[17] Interviews, February 7, 2013, and June 24, 2013.

difficult to find opportunities to centralize assessment data to build a more comprehensive picture of the AO.[18]

Traditional rivalries aside, the U.S. military at all echelons should take greater pains to establish a working relationship with counterparts in their AOs during the earliest stages of the planning process so that priorities and associated strategy, tactics, and assessment methods can align with, be accepted by, and be interwoven into the broader interagency process. Careful attention should also be given to prioritizing metrics that capture the local context or the nature of the conflict, rather than assessing objectives that serve only to advance organizational agendas.[19]

Prior agreement on goals and assessment metrics (driven by and directly linked to the commander's campaign objectives) can also serve as a "forcing function" to help in getting people to aim at the same target and ensure unity of effort. This might start with an interagency/international working group in which each member is tasked to leverage his or her specific expertise (relevant databases, SMEs, personal experiences) to identify a set of variables across all lines of effort (security, governance, development). The goal would be to develop an off-the-shelf assessment capability that uses a few "standard" metrics accepted by the broader community but also allows room for customization, depending on the specifics of the OE.[20] It is important to note that getting all the right people in the room, not to mention getting them all to agree on a common operating picture, is an ambitious endeavor that could actually stymie progress rather than facilitate it. But even if such planning events do not produce an agreed framework, they can serve as opportunities to identify and understand how other stakeholders view stability and the objectives for achieving it.[21]

---

[18] Interview, March 7, 2013.

[19] Comments of external reviewer, November 5, 2013.

[20] Interview, February 12, 2013; Becker and Grossman-Vermaas, 2011.

[21] Based on author's experience collaborating with coalition forces, the interagency, and NGO participants in Afghanistan in 2009.

Challenge 4: The Strategic- vs. Tactical-Level Challenge: Too Much Aggregation Obfuscates Nuance, Too Little Can Overwhelm Consumers

> Aggregation isn't just hard; it isn't the right way to think about moving from one level of analysis to the next.[22]

Findings

Commanders at the theater level have precious little time to sift through the nuances of every event affecting every village within their area of responsibility. Attempting to do so would obscure the broader understanding of the campaign's progress. Analysts therefore often aggregate large amounts of data as a way for commanders to consume an otherwise indigestible avalanche of information. But how much aggregation is too much? Rolling up metrics into higher and higher composites can be viewed as misleading because it strips out the operational context of events. For example, an increase in the number of direct-fire engagements is almost inevitably read as negative when viewed at the theater level, but at the small-unit level, it may be an artifact of commanders going into a deliberate clearing phase of their local COIN effort and a sign that they are seizing the initiative from the insurgents—a net positive (Connable, 2012; Schroden, 2009). The level of detail required at the theater level would be utterly insufficient for commanders at the division or battalion level, so the art is identifying the right level of detail and context for each unit echelon.

Further complicating the issue, it is not certain that one can or should aggregate the same metrics from different locations and divergent unit objectives. For example, a unit could be conducting a locally focused stability operation that is perfectly successful in terms of linking operations to sensible local stability objectives and identifying appropriate metrics to measure effects. Yet applying this same operational approach beyond that localized area/conflict could exacerbate political tensions and other problems elsewhere in the district.[23] Put

[22] Interview, March 1, 2013.

[23] Interview, March 1, 2013.

another way, the theory of victory at the local level may be different from the theory of victory at the national or regional level. Essentially, the whole may not simply be the sum of its parts.

Critics also contend that commanders have developed an unhealthy obsession with "coloring-book assessments" (Upshur, Roginski, and Kilcullen, 2012). This refers to the widely used red-yellow-green stoplight charts that may be valuable for giving an impressionistic view of the OE but rarely provide a sufficient understanding of the nuances of an area or the metrics underpinning the data to make operational decisions. As one critic stated:

> Regional commands appear to be "color averaging" when attempting to combine assessments from separate lines of operation. The regional commands present separate colors for their respective regions for security, governance, and development, then provide an overall assessment color that happens to be the average point on the color-bar chart of the three lines of operation. This is not coincidence; I have observed regional command briefers struggle to explain in operational terms why they had given a particular color to an overall assessment. (Downes-Martin, 2011)

Additionally, quantitative assessments seldom capture the whole picture. Reporting requirements often focus on "shuras [community councils with varying levels of formality, typically populated by male village elders] held versus local goodwill, number of partnered operations rather than real relationships built outside the wire, dollars spent versus actual popular commitment, IEDs [improvised explosive devices] found versus demonstrated local security forces' readiness" (Cancian, 2013). In one case, a unit was using the number of schools built as an indicator for community stabilization in Afghanistan and later found that the Taliban was teaching classes in some of those schools.[24] Narrative is key to providing context to numbers.[25]

---

[24] Interview, February 7, 2013.

[25] Interview, February 15, 2013.

Recommendations

While aggregation has drawbacks, all assessments include summation to some degree. The key is to present the results in a way that can support more detailed interrogation/exploration of the data should there be a need for further clarification. One SME noted that incorporating a sanity check by triangulating data prior to aggregation at higher levels may increase the value of the information: "It will be more qualitative than quantitative but will more closely match what frustrated commanders do currently; ask their subordinate commander what they think."[26]

Striking the right balance between quantitative and qualitative data should be a priority concern for all those involved in assessment. Both strategies have their strengths and weaknesses; the art is to use them in complementary fashion. The question of whether they are useful or correct is also decided by how one links them back to the objectives they are supposedly meeting (Stewart, 2013). Unfortunately, as the next section demonstrates, establishing a causal link between objectives and effects is rarely a simple task.

Challenge 5: Assessments Sometimes Rely on Invalid or Untested Assumptions About Causes and Effects

Findings

Complex operating environments make it difficult to establish cause and effect when assessing the impact of stability operations. For example, on its face, building a dam that provides water for the crops of a local village seems like a good thing to do and a positive step toward creating stability. Yet predicting the second- and third-order effects of the activity (e.g., an angry neighboring village whose water source has now been depleted) can be tricky (U.S. Joint Chiefs of Staff, 2011c). Even seemingly simple metrics such as SIGACT counts can be interpreted in different ways: Does an increasing number of attacks after the start of LFSO mean the mission is failing, or does it mean that the insurgents are desperately trying to slow down a winning strategy?[27]

---

[26] Comments by an external reviewer, November 2013.

[27] Interview, February 14, 2013.

Is an area with limited violence securely under government control, or is it simply quiet because U.S., HN, or other opposing forces are not present? These examples not only illustrate the importance of adding qualitative data to an assessment, they also underscore the importance of understanding the causal relationship between variables in order to validate assumptions one has about different dynamics in the operating environment. Table 3.1 provides additional examples.

Recommendations

Commanders and staffs must be wary of drawing hasty conclusions in their efforts to establish local security and should continuously identify/document and test/validate assumptions and adjust the Theory of Change (discussed below) accordingly.

Table 3.1
Assumptions and Potential Alternative Causal Effects or Interpretations

Assumption	Potential Alternative Effect or Interpretation
Development leads to stability.	But . . . it may also lead to increased competition/conflict over the resources made available by development.
Reduction in violence is a positive development.	But . . . it may also indicate complete control of an area by the opposing force.
Roads provide access to markets and health care.	But . . . they may also increase freedom of movement for the opposing force.
Stability must precede development.	But . . . increased development may lead to the desired increase in stability.
Increase in security capacity will reduce violence.	But . . . it can also lead to increased suppression and corruption.
Long-term presence is a key to success.	But . . . it can also foster resentment and/or create dependencies.

Challenge 6: Bias, Conflicts of Interest, and Other External Factors Can Create Perverse Incentives

> Sure, when doing holistic assessment, you are naturally going to look toward the data that show your success.[28]

Conflicts of interest and other external factors can create perverse incentives that muddy the assessment waters. The 2014 withdrawal date for coalition forces has led to a series of unintended consequences with respect to objectively assessing Afghan security-force capacity. Several interviewees noted that the rush for the door has created incentives to produce inflated reports of the operational readiness of Afghan units.[29] In one security initiative, stability—or at least transition in some areas—was assessed more by externalities (such as the end of a private-security-firm contract) than by actual HN performance.[30] Additionally, critics note that commanders may have an incentive to spin data, as poor HN performance is seen as a reflection of the commander's inability to achieve mission objectives (Cordesman, Mausner, and Lemieux, 2010).

Inexperienced assessment teams might also be unaware of inherent biases that can influence an assessment instrument. For example, polling is a crucial assessment tool for measuring stability, yet it is fraught with problems. Sample bias may lead to data that inaccurately represent the population (the polling population may overrepresent a particular tribe, thus skewing the data). Locals may be reluctant to answer questions honestly for fear of reprisal (from corrupt HN security forces and/or insurgents in the area) or because of a perception of

---

[28] Interview, February 14, 2013.

[29] LaRivee (2011) identified the risk that groups or individuals might manipulate metrics to reflect the signals the subjects of oversight want to send rather than the reality of the current condition. These "captured metrics" may be used to promote agendas and can provide misleading information on the effectiveness of governance or local forces or can appear to negate assumptions regarding the relationships between COIN activities and their effects on key conditions.

[30] Meeting with the authors, January 25, 2013. This was not a consented interview but observations noted during a RAND project meeting on a related topic.

what is "correct" or socially acceptable to say, rather than an accurate reflection of their views (Eles et al., 2012). The manner in which analysts conduct the survey (solo interviews vs. a group setting, phone interviews vs. in person) also affects whether and how villagers might respond.

Survey wording may alienate the population surveyed, causing respondents to lie or refrain from answering questions altogether. In addition, an interviewer's obvious association with a combatant organization affects the openness and honesty of respondents, as does the power disparity between a member of an occupying military force and an unarmed local resident. Cultural practices may also discourage public and open criticism of state institutions. In a September 2011 poll, 52 percent of Afghani respondents reported that they felt uncomfortable about publicly criticizing the government of Afghanistan (Eles et al., 2012).

Recommendations

When in doubt, assessment analysts should use additional sources for checks and balances.[31] For example, "devil's advocacy" should be incorporated into the assessment process. Devil's advocacy allows analysts to establish bounds by setting up an adversarial process: What is the worst possible assessment we can give and what is the best possible assessment? Now, where do we really think we fall between those two?[32] Devil's advocacy not only looks at things in the worst possible light, it provides contrarian views that highlight alternative Theories of Change or alternative interpretations of the data.

Assessment teams should also devote significant efforts to understanding the cultural context of the sampling environment in order to design, implement, and accurately interpret the data from population surveys. Remotely observable behavioral indicators are useful (observing behavior at the bazaar or whether children are attending schools, rather than simply asking people if they feel "safe" in their community). In addition, assessment teams should make much more effective

---

[31] Interview, January 8, 2013.

[32] Interview, February 12, 2013.

use of other sources of information. One practical strategy for developing assessments is to use local mechanisms for gathering information. In Afghanistan, this might mean exploiting shuras to understand the sources of conflict and instability in the community. It may be necessary to recruit indigenous sources to find out what is being said within the shura and in the community more broadly. U.S. forces should also make broader use of anthropologists or other SMEs who can help "map" the social terrain.[33] They should also consider partnering with local security and intelligence organizations (e.g., the National Directorate of Security in Afghanistan), while recognizing that they have different agendas that will bias their inputs in particular ways.[34]

It is important to keep in mind that data from any source run the risk of being skewed by organizations that have hidden agendas and preferred outcomes.[35] Interviewees offered numerous examples from experiences in Africa and Afghanistan in which NGOs had published scathing reports on how the military was conducting stability operations, based on the NGOs' survey work in an area, but had skewed the questions to favor their own presupposed views. Some (though not all) of the reporting was found to be based more on misleading data that served the purposes and goals of the organization than on the truth on the ground.[36] Even independent polling companies have incentives to skew their results to match the expectations of their clients.[37]

We have addressed some of the challenges associated with biases and potential ways to overcome them.[38] Though never perfect, such approaches provide the assessor a chance to "triangulate to validate"

[33] Interviews, February 7, 2013, and June 24, 2013.

[34] Interview, February 7, 2013.

[35] Upshur, Roginski, and Kilcullen, 2012; interview, February 22, 2013.

[36] Interviews, February 22, 2013, and March 7, 2013.

[37] Comment by an assessment SME with experience as a deployed analyst in Afghanistan, November 2013.

[38] A more detailed discussion is provided in a forthcoming RAND report by Dan Madden.

by using multiple methods to confirm otherwise potentially dubious data.[39]

Challenge 7: Redundant Reporting Requirements and Knowledge-Management Challenges Impede the Assessment Process

> When he got there, [the previous unit] had 230 metrics, many of them unobservable. They were drowning in metrics.[40]

Findings

Units in the field are under constant operational pressures that force them to carefully prioritize how they spend their time. When one Special Operations Forces officer was asked how he assessed progress toward stabilizing his area of responsibility in Afghanistan, he recalled that he was "just trying to survive the day."[41] Tactical-level operators expressed frustration with the cumbersome and overly burdensome amount of the reporting requirements thrust upon them. One interviewee mentioned a reporting requirement that included 900 variables. And this was not entirely an anomaly—several others mentioned filling out questionnaires with lists of questions that went well into the hundreds.[42]

Recommendations

To reduce the burden on units, it may be more efficient to ask for data less frequently but require more in-depth responses or ask for data more often but use less onerous questions.[43] An assessment team should weigh the benefits and costs of relaxing the reporting requirement. Reducing the requirement might allow for more quality reporting from teams

---

[39] Interview, February 14, 2013.

[40] Interview, February 14, 2013.

[41] Interview, February 15, 2013.

[42] Meeting with the authors, January 25, 2013. This was not a consented interview, but relevant insights the authors wrote down during a project meeting; interview, February 15, 2013; Upshur, Roginski, and Kilcullen, 2012.

[43] Interview, February 22, 2013.

that are not burned out or overly burdened, but a reduced reporting schedule might risk losing important data collection opportunities due to personnel forgetting what had occurred over a month's time rather than two weeks.[44] Determining which method should be implemented is beyond the scope of this study. However, if afforded the opportunity, analysts should experiment with different data-collection times and depth to ensure quality reporting.

One reviewer suggested integrating the concept of sending/embedding assessment personnel down to field units or to those offices/entities that have the assessment data required. This could lighten the RFI burden and ensure that the right information gets to the assessment team.[45] However, facilitating the visit of a survey team creates work for the hosting unit, and care must be taken (organizationally and in terms of personality) to avoid giving the impression that the hosting unit is being put under a microscope.

It is also useful for analysts to explain to their counterparts the purpose and process of an assessment in a meaningful way. An Operational Detachment Alpha captain said that in many cases, his team felt that it was reporting "just for the sake of reporting."[46] He described being tasked with a daily assessment of how capable an HN unit was. "The report would be sent up, filed away, and likely never referenced again." He concluded this based on experiences with higher headquarters personnel who, when his team conducted post-mission debriefings, always asked questions about the very same topics that had been extensively reported on, daily, for the previous few months. This further exacerbated the team's negative view of the value of conducting frequent assessments at the tactical level.

To mitigate this situation, analysts should send assessment findings back to soldiers who provide the information to show how it is being used. Adding the "why" to the "what" could increase their understanding of and enthusiasm for the reporting process.[47] As one

---

[44] Interview, February 7, 2013.

[45] Comments from external reviewer, November 5, 2013.

[46] Interview, February 6, 2013.

[47] Interview, February 22, 2013.

interviewee noted, "If you ask for a list of numbers, you'll get a list of numbers, but they may be collected inconsistently or just made up. If you ask for an explanation, an account, a reason, something connected to a hypothesis of a Theory of Change, you'll do better."[48] Sending assessment findings back to those who provide the original data can benefit both analysts and operators: Analysts get a sanity check from soldiers who can verify the results, and soldiers get tailored analytical products that may be of direct use to ongoing operations.

Several analysts interviewed strongly advocated for making greater use of operational and intelligence data that are already being collected.[49] Since the data are collected as part of a unit's core activities, response rates and accuracy are already very high relative to those of assessment-specific data. Ground Moving Target Indicator Radar and Blue Force Tracker are seen as particularly promising avenues to explore. For instance, if Ground Moving Target Indicator Radar data could be obtained for civilian areas of interest over a long enough period of time, assuming the analyst knows what to look for,[50] it might be possible to infer changes in security and economic activity and thus be better able to assess the effectiveness of coalition activities by comparing "treatment groups" with "control groups" in a quasi-natural experiment.

This recommendation sounds good in theory and could be particularly valuable in situations that require a light U.S. military footprint.[51] Yet many are skeptical that even if the information existed they would ever be able to access it. One interviewee said, "Someone has probably figured out a solution to every problem in Afghanistan at

---

[48] Interview, February 12, 2013.

[49] Interview, November 16, 2012.

[50] Interpreting the data is a challenge as well. Intelligence, surveillance, and reconnaissance capabilities have evolved over the last decade of COIN-centric warfare and could prove increasingly useful for the purposes of assessment. But without the analytical ability to understand and interpret the historical, sociocultural dynamics of the operating environment, valuable assessment opportunities are lost. For more detail, see Office of the Under Secretary of Defense for Acquisition, Technology and Logistics, 2011.

[51] Comments from an external reviewer, November 5, 2013.

some point. It's all out there but it gets lost because there's no real way to catalog or access this information in an efficient way."[52] The knowledge-management problem is tough, and neither our interviewees, nor the literature, nor our study team was able to tackle it. The topic nevertheless warrants attention and should be explored in future research. That said, no matter how willing a unit is to provide information, intentions can easily be buried under other priorities coming down the chain of command. Having the commander's support and buy-in for the data-collection schedule is a sine qua non for invigorating the assessment process and is discussed in further detail later in this chapter.

Challenge 8: Continuity of the Assessment Process Can Be Difficult Across Deployment Cycles

> In studying the DSF, I have yet to find a successful case where personnel conducting DSF handed over the assessment process to their replacements.[53]

Findings

Deployment cycles can have a negative impact on the assessment process. Depending on command guidance, newcomers may simply not be interested in what the people they replace did prior to their arrival. In some cases, new units or individuals transfer in but are provided little or no handover of relevant information/assessment processes. One SME observed that "many locals in Iraq and Afghanistan openly made fun of the USG [U.S. government] for going in and asking the same four questions over and over."[54] In the data collector's defense, personnel are sometimes following a specific methodological approach in efforts to track metrics over long periods of time. Regardless, this problem is indicative of a larger issue concerning continuity of effort and assessment.

---

[52] Interview, February 6, 2013.

[53] Comments from an external reviewer, November 5, 2013.

[54] Comments from an external reviewer, November 5, 2013.

Recommendations

Identifying ways to overcome deployment challenges is relatively easy: Shifting command guidance should come with an associated shift in the metrics expected to measure it. The RIP/TOA (relief in place/transfer of authority) process should include an explanation of how assessments are being done in the OE. Survey-fatigued respondents should be told that questions repeated over time are meant to capture trends rather than to make up for lost data. These actions could go a long way toward ensuring continuity in the assessment process. The hard part is ensuring that time and attention are paid to seeing it through.

Challenge 9: Assessment Planning Often Ignores HN Perspectives

> If we were to go through this process again, we would use more local advisors earlier on to help decide the important measures, and even what goals to pick. (Becker and Grossman-Vermaas, 2011)

Findings

Few interviewees could remember much HN participation in the assessment planning process. For example, coalition forces in Afghanistan hold copious meetings on a daily, weekly, and monthly basis, and although much of what is discussed is meant to help shape the course of the future of Afghanistan, the meetings are often entirely void of an Afghan presence.[55] One interviewee with experience as a Civil Affairs captain in Iraq offered anecdotal evidence of a U.S.-built dump site with state-of-the-art facilities and equipment that cost millions of dollars to build, but in the absence of input from HN officials, the Iraqis were scratching their heads trying to figure out its purpose and value in terms of creating stability, and even how to maintain it following the U.S. withdrawal.[56] Thus, while the building of the dump site could be assessed as a "success" at the measure-of-progress level (Level 3 in

---

[55] Author's experience as a deployed analyst in Kabul, Afghanistan, August–December 2011.

[56] Interviews, January 11, 2013, and January 15, 2013.

the hierarchy of evaluation—see Figure 3.4 below), the impact of the dump site was neutral to negative, and thus, at the MOE level (Level 4 in the hierarchy of evaluation), even a successfully built dump site would have to be assessed as a failure.

Recommendations

Assessment staff must be aware of and account for what is appropriate for the local context and relevant to the nature of the conflict while still being aligned with U.S. and HN government laws, interests, and objectives. Assessments should take into account local perceptions of what stability means rather than relying on the preconceived notions of outsiders who do not understand the complex environment. For example, a local villager may assess his village as stable even if its security is provided by insurgents. As long as he can live his life in relative peace, that villager does not care who is providing security.[57] Understanding the local environment does not mean blindly adopting the locals' goals, but having such a dialogue can inject greater realism into the design of the assessment process—and, beyond assessment, into planning at all levels of a campaign.

However, while it might be useful to involve locals in conversations about how to establish and assess security, the locals may have reasons for misleading interlopers, and even when they are being truthful, the underlying issues may remain opaque to U.S. forces because of cultural issues.[58] This does not mean discounting the locals' views altogether. Rather, there should be some "front office" discussions that are as inclusive as possible, so that planners and operators can take them into consideration during their later "back office" planning sessions.

Implementation-Related Guidance

Having laid out some of the foundational challenges facing assessment of LFSO and some related recommendations, we now discuss implementation-related guidance that can support assessment.

---

[57] Interview, February 20, 2013.

[58] Interview, February 7, 2013.

Assessment Requires Properly Defined Mission Objectives and Concept of Operations

> Objectives are often vague and diffuse, which makes assessing progress toward those objectives impossible.[59]

There can be little surprise that operational assessment processes are influenced by the ambiguity of the mission—it is hard for any metrics system to be more precise than the goals against which it is designed to measure progress (Upshur, Roginski, and Kilcullen, 2012). Further, although a unit commander knows his specific mission, the overall strategy is often unclear, making it difficult to tie local success to strategic objectives.[60] For example, when the research team asked interviewees to describe the objectives of stability operations, there were as many interpretations as there were individuals interviewed. Most of the comments seemed to center on establishing security, governance, and development, but opinions of who should do what, how much should be done, when, and why diverged as much as the methods used for assessment.

Additionally, mission objectives may change over the course of a campaign. Each commander comes into the theater with a preconceived notion of how to conduct stability operations. Since the process is based on the commander's intent, a new commander brings a new intent, with cascading changes that are not always recalibrated in the assessment framework. If mission objectives are in constant flux, consistently measuring progress is a difficult, if not meaningless, enterprise. The results tend to be processes and products that do not deliver the results commanders expect. Since commanders must weigh the costs (time, resources) against the benefits of rigorous assessment, they may curtail the assessment effort altogether, relying on subordinate command staff or others to guide their decisionmaking processes.[61]

---

[59] Interview, February 23, 2013.

[60] Interview, January 15, 2013.

[61] Interview, November 16, 2012.

To mitigate this instability, commanders should recognize and embrace their dual roles as developers and customers of assessments. They are responsible for providing guidance that can be translated into clear objectives and executed at all echelons of the mission. "The objective is key, and the objective has to come from the commanders. We need to push harder on the commanders to be clear about objectives. You want to stabilize . . . what does that mean?"[62] The commanders' guidance must also drive the assessment process. A core assessment paradigm (metrics, methods, processes, and products) aligned with the overall campaign plan should be established at the joint task force or theater level, but it should be flexible enough for local commanders to supplement it to address their own needs. Some instability will result as campaign objectives change, but the instability generated by rotations at subordinate levels in the chain of command will be minimized.

Fully Exploit the Data and Methods Available, Leveraging the Strengths of One Source Against the Weaknesses of Others to Triangulate the Truth

LFSO often take place in complex and uncertain security environments. The relatively minimal previous U.S. military presence in some areas means that data-collection and analysis teams will find it challenging to fulfill all RFIs and may be tempted to "guesstimate" when data are unavailable or too difficult to collect—especially if commanders are unwilling to accept "I don't know" for an answer (LaRivee, 2011). Commanders must understand this pressure and balance their requests against the reality of the security environment, the availability of data, and the reliability of raw information from the field (Upshur, Roginski, and Kilcullen, 2012). If requested information is not available, personnel responsible for implementing assessments should highlight (not hide) the gaps and try to adjust the collection process to obtain what is needed using proxy data. This provides a wider set of options for meeting the commander's needs while avoiding the risk that an assessment will be derived from manufactured data—although it obviously might create challenges for analysts attempting to compare

[62] Interview, February 12, 2013.

disparate data over time or across regions, complicating trend analysis (LaRivee, 2011).

An Adaptive Assessment Process Should Include Robust Input from Subordinate Commands and Staff

The operations and campaign objectives established in an operations order will not be able to dynamically capture the situational awareness or perspectives of the full community of stakeholders, so processes should be in place for staff input to update the assessment model even before the data begin to identify challenges. After-action reviews conducted after each major assessment period can help determine whether the assessment is resulting in any new command decisions. If it is not, the assessment team should determine whether the campaign is not on track or the assessment in its current form is not useful to the commander. Consistent engagement between analysts and command staff is critical to ensuring that command objectives and measures for assessing them are aligned.

In conclusion, the relationship between a commander and his assessment staff is a two-way street: The assessment process must fully support the commander, and the commander must fully support the assessment process if it is expected to provide a useful tool for shaping the campaign. Overcoming many of the foundational challenges can help in this regard. Some additional assessment principles are summarized below.

Assessments should

- Be covered in the training that commanders receive, including their purpose, utility, and how they can be effective decision-support tools (Schroden, 2011).
- Triangulate the truth by incorporating data from multiple sources (e.g., subordinate commanders; centralized reporting, such as Combined Information Data Network Exchange; SIGACTs; SMEs; polling), and methods (e.g., regression analysis, content analysis, debate among key staff and commanders).
- Include a focus on issues commanders can affect through new decisions.

- Be prioritized by the command in order to ensure that subordinates take on the task with diligence and vigor (assessment requires both top-down and bottom-up support).
- Start during campaign or mission planning.[63]
- Include input from staff, subordinate units, external centers of excellence, and other outside experts (U.S. Joint Chiefs of Staff, 2011c).
- Be sensitive to reporting-requirement fatigue of subordinates (make data requirements and collection frequencies reasonable).
- Be continually reviewed, updated, and modified as the operational environment, end state, or problem changes.
- Anticipate personnel change/rotation by documenting objectives, processes, methods, and products in order to ensure consistent objectives and measurement over time.
- Consolidate, summarize, and communicate data that best support the commander's decisionmaking in a clear, concise, and timely way (Military Operations Research Society, 2012).

A Theory of Change Should Be at the Core of Every Assessment

One assessment principle stands above others and is central in our recommended assessment framework: the importance of a Theory of Change, a concept used in project-management practice (Organizational Research Services, 2004). The Theory of Change for an activity, line of effort, or operation is the underlying logic of how the planners believe the things that will be done in the effort will lead to the desired results. Simply put, a Theory of Change describes the chain of consequences—how one believes one's actions will lead to the objectives one seeks to achieve. A Theory of Change can include assumptions, beliefs, or doctrinal principles, although none of these items in itself constitutes a Theory of Change. In campaign planning, the

[63] Interview, February 14, 2013.

Theory of Change is typically articulated through a concept of operations, although many assumptions are often left unstated.

The main benefit of articulating a Theory of Change in the assessment context is that it allows assumptions to be turned into *hypotheses* that can then be tested explicitly as part of the assessment process; any failed hypotheses can be replaced in subsequent efforts until a validated logical chain connects activities with objectives and objectives are met. This is shown graphically in Figure 3.2.

An example of an LFSO Theory of Change might be: *Training and arming local security guards makes them more willing and able to resist insurgents, which will increase security in the locale. Increased security will lead to increased perceptions of security, which will promote participation in local government, which will lead to better governance.*

Figure 3.2
A Generic Theory of Change

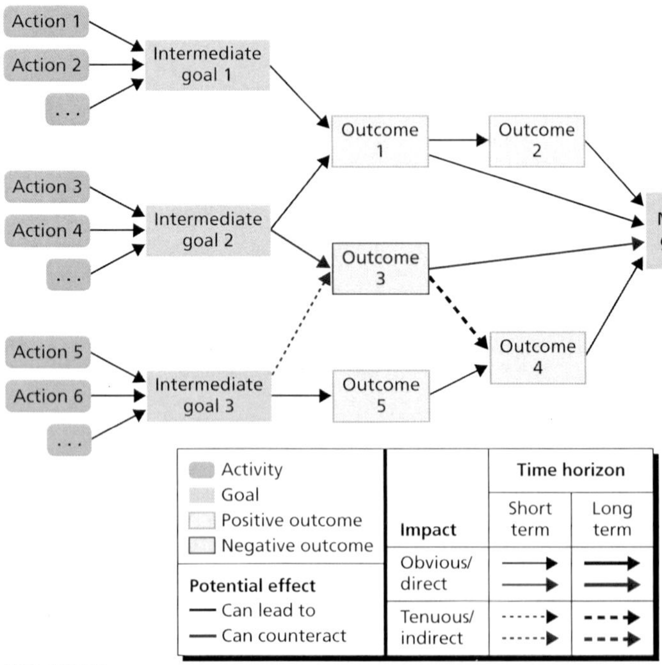

RAND *RR387-3.2*

Improved security and better governance will lead to increased stability. Figure 3.3, using the notation introduced in Figure 3.2, illustrates this and additional chains of consequences.

This Theory of Change shows a clear logical connection between the activities (training and arming locals) and the desired outcome (increased stability). It makes some assumptions about causes, as expressed by the arrows, but those assumptions are clearly stated. Further, the activities and assumptions suggest things to measure—performance of the activities (the training and arming) and outcome (change in stability)—as well as elements of all of the intermediate nodes—capability and willingness of local security forces, change in security, change in perception of security, change in participation in local government, and change in governance. Thus, if one of those measurements does not yield the desired results, it is easy to identify where the logic is breaking down and make changes to the Theory

Figure 3.3
Simplified Partial Theory of Change for LFSO

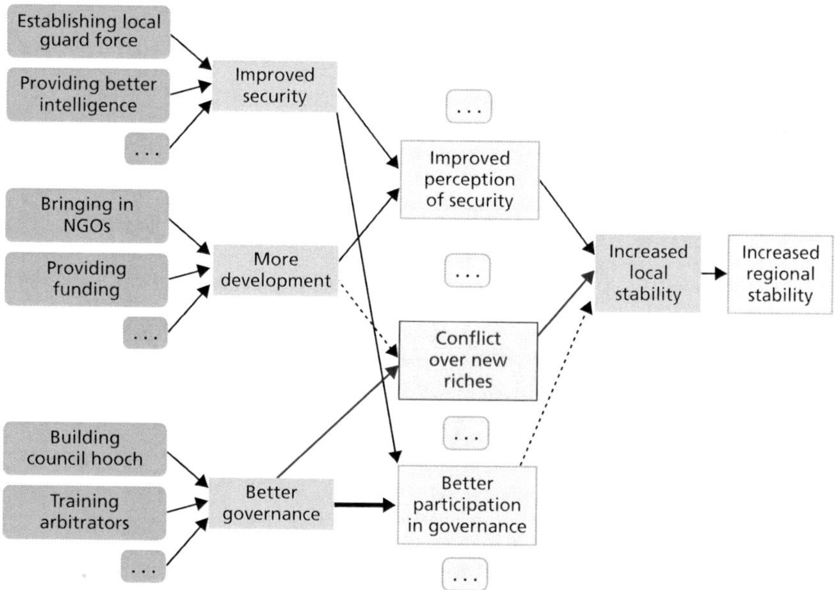

RAND *RR387-3.3*

of Change and the activities to reconnect the logical pathway and continue to push toward the objectives.

Understanding the chain of consequences that links actions to multiple outcomes also helps avoid unintended consequences of such actions—e.g., if building a new well is undertaken without taking into account local context, such as traditional power structures and water-related customs, even a successful construction that leads to an objectively improved water supply may be detrimental to stability if its location marginalizes certain factions or families in the village or if its operation leads to disruption of land-use patterns. Thus, the way something is accomplished can be as important as what is being accomplished.

Iteration Is Important

A Theory of Change may begin with something quite simple, such as *training and arming local security guards will lead to increased stability.* While this captures the kernel of the theory, it is not particularly helpful, as it suggests that we need to measure only the activity and the outcome. It leaves a huge assumptive gap: If training and arming goes well, but stability does not increase, we will have no idea why. To begin to expand on a simple Theory of Change, we ask the question, Why? How do you think that A leads to B? (In this example, How do you think that training and arming leads to stability?) A thoughtful answer usually adds another node to the Theory of Change. The question can be asked again relative to this new node until the Theory of Change is sufficiently articulated.

There is no hard and fast rule for knowing when the Theory of Change is complete; this is at least as much art as it is science. Too many nodes and too much detail can result in something like the infamous Afghan Stability/COIN Dynamics "spaghetti diagram" that captures most causal relationships but is too complex and chaotic to grasp (PA Consulting Group, 2009, slide 22). Adding too few nodes results in something too simple that leaves too many assumptive gaps. If an added node evokes thoughts like "Well, that is obvious," perhaps it is overly detailed.

If an initial Theory of Change is not sufficiently detailed, iterative assessments will point toward places where more detail is required. A measurement that is positive on one side of a node but negative on the other suggests either that there is a mistaken assumption or that an additional node is required. For example, measures may show real increases in security on one side—reduced SIGACTs, reduced total number of attacks/incursions, reduced casualties and/or cost per attack—but measures of perception of security, from surveys and focus groups, as well as observed market and street presences, may not correspond. If the team is unwilling to give up the assumption that improvements in security lead to improvements in perceptions of security, it will need to look for another node. The team can speculate and add another node or do some quick data-collection activities, developing a hypothesis from operators or from local special focus groups. If the missing node *awareness of the changing security situation* is confirmed by preliminary information, the assessment team may need to add an additional node to the Theory of Change, as well as an additional factor to measure. A new activity may also have to take place, such as some kind of effort to increase the awareness of changes in the security situation.

Improvements to the Theory of Change not only enhance assessments, they can improve operations. Articulating a Theory of Change also allows assessment teams to begin activities with some questionable assumptions, knowing that they will then be either validated by assessment or revised. Theory of Change–based assessments support learning and adapting throughout operations.

A preliminary Theory of Change can be improved by asking after connective nodes, How do you think A leads to B? It can also be improved by asking after missing disruptive nodes, What might prevent A from leading to B? or What could disrupt the connection between A and B? Such questions can help identify possible logical spoilers or disruptive factors. Again, if the possible disrupter is articulated as part of the Theory of Change, it is relatively easy to then seek to measure it. For example, when connecting training and arming local guards to improved willingness and capability to resist insurgents, we might add to the Red Team disruptive factors such as "trained and armed locals defect to the insurgency" or "local guards sell weapons instead of keeping them." We can then add these spoilers to the Theory of Change (red text and arrows in Figures 3.2 and 3.3) and seek to measure the possible presence of them.

Another way to ensure that a Theory of Change is sufficiently articulated is to verify that it has nodes (and accompanying measures) at a minimum of three levels—Level 2, Level 3, and Level 4—in the hierarchy of evaluation (discussed below).

The Hierarchy of Evaluation

The hierarchy of evaluation developed by Rossi, Lipsey, and Freeman (2004) is presented in Figure 3.4. It divides all potential evaluations and assessments into five levels that are nested such that each higher level is predicated on success at a lower level. For example, positive assessments of cost-effectiveness (the highest level) are possible only if they are supported by positive assessments at all other levels. Further details are given below in the subsection on hierarchy and nesting.

Figure 3.4
The Hierarchy of Evaluation

SOURCE: Adapted from Figure 7.1 in Paul et al., 2006, p. 110.
RAND *RR387-3.4*

Level 1: Assessment of Need for Effort

Level 1 is the assessment of the need for the program or activity. This foundation is where evaluation connects most explicitly with target ends or goals. It focuses on the problem to be solved or goal to be met, the population to be served, and the kinds of services that might contribute to a solution (Rossi, Lipsey, and Freeman, 2004, p. 76).

Evaluation at the needs-assessment level is often skipped, being regarded as wholly obvious or assumed. For situations in which such a need is genuinely obvious or the policy assumptions are good, this is not problematic. But where need is not obvious or goals are not well articulated, troubles starting at Level 1 can complicate assessment at each higher level.

Level 2: Assessment of Design and Theory

Once the needs assessment of Level 1 establishes that there is a problem or policy goal to pursue and identifies the intended objectives of such a policy, different solutions can be considered. Assessment at Level 2 focuses on the design of a policy or program; this is the level at which an explicit Theory of Change should be articulated and assessed. It is also a critical and foundational level in the hierarchy. Unfortunately, this level of evaluation also is often skipped or completed minimally and based on unfounded assumptions.

Level 3: Assessment of Process and Implementation

Level 3 focuses on program operations and the execution of the elements in Level 2. Efforts can be perfectly executed but still not achieve their goals if the design was inadequate. Conversely, poor execution can foil the most brilliant design. For example, a well-designed series of training exercises could fail to achieve desired results if the executing personnel did not show up or did not have the needed equipment. Level 3 evaluations include "outputs," the countable deliverables of a program. Traditional measurements at Level 3 are measures of performance.

Level 4: Assessment of Outcome/Impact

Level 4 is near the top of the evaluation hierarchy and includes outcomes and impact. At this level, outputs are translated into outcomes, a level of performance, or achievement. Put another way, *outputs* are the products of activities, and *outcomes* are the changes resulting from these activities. This is the first level of assessment at which solutions to the problem that originally motivated efforts can be seen. Measures at this level are often referred to as MOEs.

Level 5: Assessment of Cost-Effectiveness

The assessment of cost-effectiveness is at the top of the evaluation hierarchy. Only when desired outcomes are at least partially observed can efforts be made to assess their cost-effectiveness. Simply stated, before you can measure "bang for the buck," you have to be able to measure the "bang."

Evaluations at this level are often most attractive in bottom-line terms, but they depend heavily on lower levels of evaluation. Measuring cost-effectiveness in situations with unclear resource flows or where exogenous factors significantly affect outcomes is complicated. As the highest level of evaluation, this assessment depends on the lower levels and can provide feedback inputs for policy decisions based primarily on those lower levels. For example, if target levels of cost-effectiveness are not being met, cost data (Level 5) in conjunction with process data (Level 3) can be used to streamline the process or otherwise selectively reduce costs.

Hierarchy and Nesting

This framework is a hierarchy because the levels are nested—solutions to problems observed at higher levels of assessment often lie at lower levels. If the desired outcomes (Level 4) are achieved at the desired levels of cost-effectiveness (Level 5), lower levels of evaluation are irrelevant. However, when desired high-level outcomes are not achieved, information from the lower levels must be available to be examined.

Thus, assessment schemes must include evaluations at a sufficiently low level to inform effective policy decisions and to enable problems to be diagnosed when the program does not perform as intended.

Conclusion

This chapter has identified some foundational challenges of assessment, and it provides implementation-related guidance for developing and executing an assessment plan. It also describes the Theory of Change and how it can be used to identify the preconditions necessary to achieve mission objectives and to document the associated assumptions. While many of these insights can be found scattered throughout the literature and in the diversity of opinions of experts on the subject, we hope that consolidating and summarizing them in a more digestible format will prove useful to those involved in the assessment process. Tables 3.2 and 3.3 summarize this collection of challenges and best practices. Chapter Four presents an illustrative application of our recommendations to a scenario based on a notional country in the AFRICOM area of responsibility.

Table 3.2
Summary of Challenges and Recommendations

Foundational Challenges	Recommendations
Assessing the impact of stability operations in a complex environment is not easy.	Identify the challenges, take a deep breath, and forge ahead.
Doctrine and training fail to adequately address complexities and appropriate skill sets.	Prioritize assessment-related doctrine/ training. Institutionalize the assessor role. Assign individuals with the "right" personality traits. Elicit SMEs to fill in contextual gaps. Allow CONUS analysts to deploy to gain operational grounding.
Visions of stability among stakeholders (United States, coalition, HN, NGOs, etc.) are often in competition.	Establish an interagency/international working group to identify a set of variables across all lines of effort (security, governance, and development). Develop an off-the-shelf assessment capability that uses a standard framework and is accepted by the broader stakeholder community.

Table 3.2—Continued

Foundational Challenges	Recommendations
A balance must be found between strategic- and tactical-level aggregation: too much aggregation obfuscates nuance; too little can overwhelm consumers.	Present results in a way that efficiently and clearly summarizes but can support more detailed exploration of data should the need arise.
Assessments sometimes rely on invalid or untested assumptions about causes and effects.	Avoid drawing hasty conclusions by identifying/documenting and testing/validating assumptions. Adjust the Theory of Change accordingly.
Bias, conflicts of interest, and other external factors can create perverse incentives.	Triangulate to validate, using observable indicators, devil's advocacy, ratios, and other multiple-sourced methods.
Redundant reporting requirements and knowledge-management challenges impede the assessment process.	Ask for data less frequently but require more in-depth response, or ask for data more often but use less onerous questions. Provide direct benefits (e.g., tailored products) to those who provide data to validate their efforts and motivate them.
Continuity of the assessment process can be difficult across deployment cycles.	Plan for personnel turnover (training, documentation, knowledge management).
Assessment planning often ignores HN perspectives.	Invite HN participation in assessment planning and execution. Carefully consider hidden agendas.

Table 3.3
Summary of Implementation-Related Guidance

Implementation-Related Guidance	Theory of Change
• Good assessment requires trained assessment staff and "assessment-aware" commanders. • Effective assessment starts during campaign or mission planning. • Developing the necessary Theory of Change benefits overall campaign/mission planning. • Effective assessment requires clearly defined objectives and concepts of operation. • Assessment should favor real (but perhaps messy) data over clean (but inaccurate) data. • The assessment process should be developed on the basis of robust input from subordinate commands and staff. • All personnel involved should be clear on the purpose of the assessment. • Assessments may focus on a project/program (local level, limited scope) or a campaign (theater level, overall effort). • Assessments may evaluate needs or baseline (focus on starting conditions) or progress (focus on delta) or may be made against an absolute standard. • Assessment planning must include how best to communicate results to decisionmakers.	• Is at the core of a successful assessment effort. • Connects activities, desired outcomes (effects), and superordinate objectives through a "Chain of Consequences." • Illustrates how outputs at a given level may become inputs at a higher level. • Can make complex scenarios more manageable. • Turns assumptions into explicit, testable hypotheses. • Helps select valid measures and indicators. • Is important for proper overall campaign design. • Has key elements shown in Figure 3.2.

CHAPTER FOUR

Applying the Assessment Recommendations in a Practical Scenario

To demonstrate the practical application of the assessment recommendations outlined in Chapter Three, we developed a comprehensive LFSO scenario based on a notional African country. This chapter describes why we chose an LFSO scenario in Africa, how we designed the context, the resulting concept of operations, and the assessment plan.

Future Prospects for Locally Focused Stability Operations in Africa

In the future, the U.S. military will likely have to focus on global trouble spots, using "a series of persistent, targeted efforts to dismantle specific networks of violent extremists that threaten America" ("Remarks by the President at the National Defense University," 2013). This approach will require partnering with countries in areas that risk becoming breeding grounds for extremism and tyranny, to build indigenous security capacity and also to help address the root causes of extremism by improving governance and supporting efforts to expand economic opportunities. Part of that strategy entails "creating reservoirs of goodwill that marginalize extremists" in areas where widespread poverty, corruption, police abuse, and the lack of government control could create fertile ground for militant extremism to take root.

While some countries in Africa are viewed as success stories because of their peaceful transition to democracy and increasing economic development, many are still challenged by rampant corruption, ethnic conflict, poverty, and Islamic militancy. The United States provides a wide range of bilateral assistance aimed at improving security capacity, strengthening governance, and increasing economic development opportunities in these countries. According to the U.S. Department of State, "U.S. assistance also aims to reinforce local and national systems; build institutional capacity in the provision of health and education services; and support improvements in agricultural productivity, job expansion in the rural sector, and increased supplies of clean energy" (U.S. Department of State, 2013). The U.S. Army can play an important role in achieving some of these objectives through security-force assistance programs and small-footprint deployments, some of which may take the form of LFSO.

Scenario for Application of Locally Focused Stability Operations

To make the context for this scenario as realistic as possible, we examined the following factors when selecting the country and conflict:

- Development level and stability of HN government
- HN security-force capabilities and capacity
- Insurgent (INS) capabilities compared with those of HN
- Level of violence
- Duration of conflict
- INS motivation
- Cross-border issues
- Traditions of local governance
- Relevance of conflict and country to U.S. national security.

Other factors may also play a role in determining the suitability of a conflict for application of LFSO, and thus the process used for such determination is a promising area for future research. The following country and conflict overview is based on a real-world example that was identified through this process and provides a comprehensive foundation for the discussion of LFSO assessment.

Country Overview of the Notional Host Nation

In 2015, a large West African country with strategic importance to the United States decides to implement targeted LFSO to overcome a number of formidable challenges it faces as a nascent democracy, including terrorist activities, sectarian conflicts, and public mistrust of the government. The central government is presumed to be relatively stable; however, some claim that corruption, rapidly increasing unemployment, and violence associated with Islamic insurgency have the potential to destabilize it. As the country is a major foreign source of oil, U.S. strategic interests are at stake. More importantly, the country's ungoverned spaces run the risk of becoming safe havens for terrorists fleeing tumultuous Northern Africa.

With a population of over 100 million and a relatively powerful economy, this HN is a political and economic powerhouse in West Africa, yet it faces a number of obstacles that threaten its stability. It is estimated that 30 percent of the working population are unemployed. The country's economic growth has been largely fueled by oil revenue, yet the Organisation for Economic Co-operation and Development considers it to be a lower-middle-income, fragile state because of its overreliance on oil and gas revenue (about 90 percent of its exports are fuel products).

Most of the country's revenue streams into the hands of its ruling elite. Illiteracy stands at 40 percent, and the percentage of poverty is rising, with more than half of the population now living in absolute poverty on less than $1 a day. Despite petroleum income, citizens have to cover all their basic services themselves. This largely extends to security as well, as the HN's security forces struggle with rampant corruption and have been accused of frequent human-rights violations, including extrajudicial killings, torture, arbitrary arrests, and extortion-related abuses. An influential NGO operating in the country observed that the continued failure of the government to address the widespread poverty, corruption, police abuse, and long-standing impunity for a range of crimes has created a fertile ground for violent militancy. Since the end of military rule more than a decade ago, more than 10,000 people have died in intercommunal, political, and sectarian violence.

The government has designated broad-spectrum changes for its security forces as part of a wider policy move to promote democratic principles, focusing on improving salaries and living and training conditions for military personnel and eliminating corrupt practices. The military in particular is undergoing a transformation aimed primarily at fostering greater efficiency and professionalism.

Despite these challenges, the HN army has demonstrated its capability in a number of areas, including fielding battalions in support of regional peacekeeping operations. HN armed forces are also engaged in border protection. Border issues are less about regional conflict than about government security forces' inability to prevent transnational organized-crime groups from using local border communities as hubs for illicit activity, terrorist camps, and gateways for trafficking drugs and weapons. Hundreds of illegal footpaths crisscross the country's borders, are mostly unknown to security agencies, and are thus unprotected and serve as transit routes to neighboring countries extending to Northern Africa.

State of the Insurgency

Of increasing concern are HN security-force COIN efforts against a burgeoning Islamic insurgency that is active in the northern region, where the president says his forces have lost control. Insurgent-related violence first erupted several years ago. Recently, insurgents have taken to murdering and intimidating civilians and local farmers, causing many to flee villages and abandon large swaths of crops during the harvest—all of which has aroused fear of national-level food shortages. Like other such movements, the insurgents reject modern narratives and seek to apply what they view as traditional religious answers to social questions.

As a political movement, the group is apparently split into at least three factions, none of which is afraid to use violence to achieve its aims. Their goals are both long-term and immediate. They have a well-developed domestic bomb-making capability, as the frequency of deadly explosions and the discovery of bomb factories demonstrate. The decentralized structure of the insurgency allows for multiple cells to conduct influence operations across an expanse of territory, focusing their efforts at the village level, where government services are often lacking or absent and resentment is high. Like other political and armed movements that have sprung up in the country, including recent fuel-subsidy protests that brought it to a standstill, the INS group's ability to manifest is just a symptom of the crumbling state. Despite their daily trials, the vast majority of locals do not turn to armed militancy, but the fact that a small and very deadly portion do is a clear sign of the country's underlying dysfunction.

The Theory of Change Ties Actions to Mission Objectives

Establishing the Theory of Change behind the planned operation helps document the expected results and describe how activities and tasks are linked to those results. The primary objective for the HN LFSO team is to stabilize the village and ensure that stability is sustainable. This, in turn, should help protect the village from INS attacks or influence operations. Three essential building blocks, which will be discussed in detail in the next section, serve as a foundation and catalyst for stability:

- Security for the village and its immediate vicinity
- Economic development geared toward increasing employment and standard of living
- Improvements to governance that link locals and the central government.

Achieving success for each of these three building blocks requires specific actions to be taken and specific preconditions to be met. Establishing security for the village will be predicated on the availability of qualified and able-bodied individuals and on the ability to train, equip, and supervise them. The resulting increase of security in and around the village should prevent farmers and locals from fleeing and should convince merchants that they can safely continue their work. Economic improvements may require changes in economic policies, subsidies for building shops, increased reliability of electrical and cell-phone networks, etc. Success in this area can result in increased economic growth, with further benefits derived from the linkages between security and economic prosperity. Finally, improved governance may require integrating local power brokers, providing a forum for collaboration, and creating lines of communication between the village and the central government. Establishing such a collaborative, inclusive political process will convince villagers that participatory politics, not *sharia* law, promotes justice and transparency, and this may lead to increased trust between villagers and government officials.

At its most simple, the Theory of Change for LFSO can be visualized as shown in Figure 4.1. LFSO efforts aimed at immediate goals in the areas of security, development, and governance lead to the main goal: increased, sustainable stability.

However, to be of operational use, more detail about driving actions and the higher-level context of the LFSO effort must be added, as illustrated in Figure 4.2. Achieving each of the three immediate goals depends on successful LFSO actions in the associated lines of operation: recruiting, training and equipping a local guard force,

Figure 4.1
Basic Theory-of-Change Chart for LFSO

RAND *RR387-4.1*

Figure 4.2
Actions Linked to Ultimate Outcomes in a More Detailed Theory-of-Change Chart

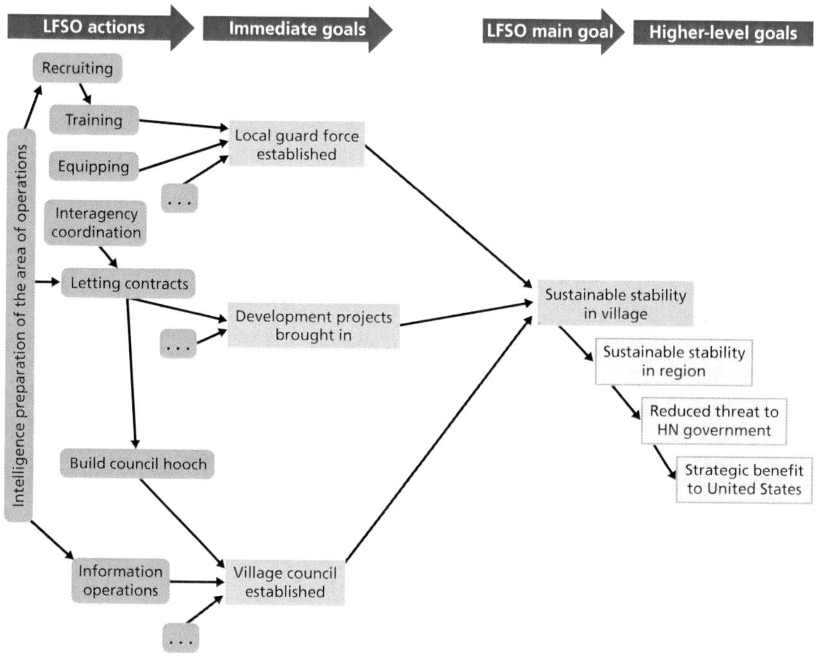

RAND *RR387-4.2*

coordinating and letting contracts for development projects in the village, and laying the groundwork for starting a village council. At the higher level, the Theory of Change now documents the assumption that sustainable stability at the village level will lead to more stability in the region, which will reduce the threat to the HN government, which, in turn, will yield the desired strategic benefit to the United States.

To flesh out the initial Theory of Change, details about the (postulated) chain of consequences leading from initial outcomes to overall objectives are added (Figure 4.3). Improvements in the immediate goals lead to outcomes such as improved security, an improved economic situation, and more effective local governance, which in turn lead to further improvements, such as reduced unemployment, reduced

Figure 4.3
Basic Chain of Consequences Added to Flesh Out Links Between Actions and Goals

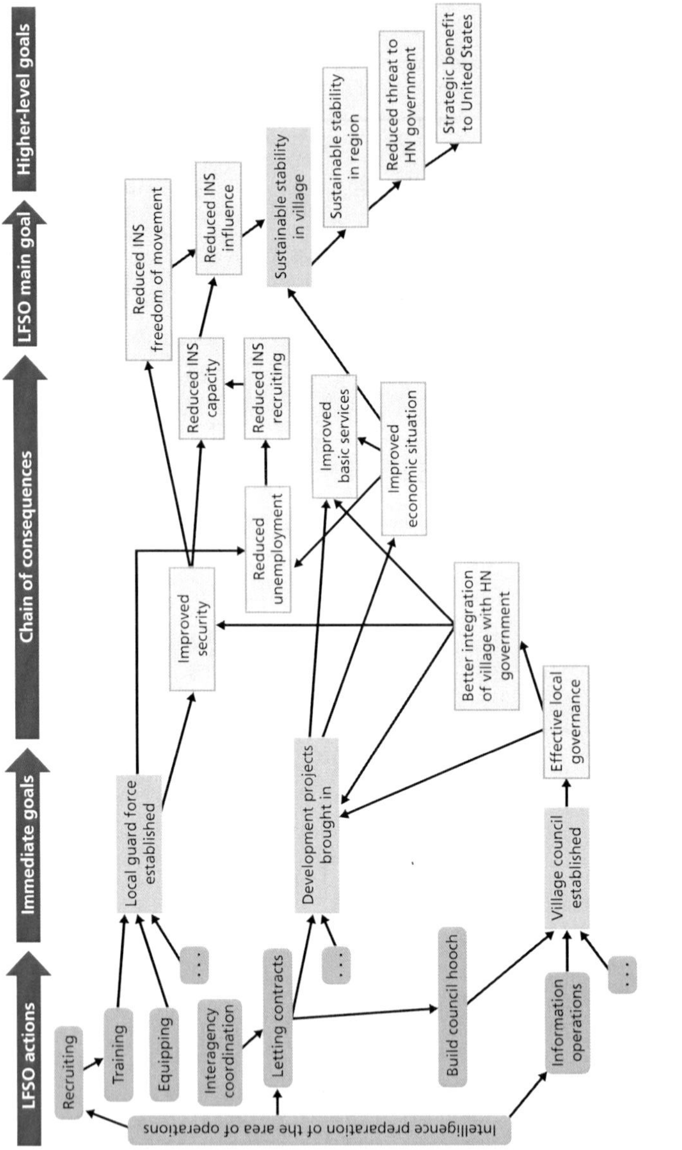

RAND *RR387-4.3*

potential for INS recruiting, and reduced INS freedom of movement. All of these effects ultimately support[1] the main goal.

Concept of Operations

Recognizing the potential for civil unrest and regional conflicts to quickly become intractable, the HN wants to implement LFSO in the north. Initial plans envision the stability operation lasting for several years, with embedded units having one-year rotations. However, to ensure success and prevent human-rights abuses that often stem from military-only stability operations, the HN plans to field army and police units that are strategically augmented by civilian government personnel, preferably from the north, with the aim of avoiding ethnic tensions and enhancing cultural capabilities. Additionally, the HN will arrange for an independent NGO to conduct village opinion polling.

While no direct U.S. involvement is foreseen at the village level, the HN LFSO teams will be trained and advised by a U.S. interagency team that works through the HN's U.S. Embassy (USEMB) and consists of personnel from the Department of State, USAID, and United States Army Africa. Importantly, the HN has also asked to have the U.S. interagency team perform a continuous independent assessment of the LFSO effort. Finally, the HN and the United States will jointly provide intelligence support to the HN LFSO team.

Following the Theory of Change outlined above, as well as lessons learned from other LFSO, such as the one in Mali in 2011 (the Special Programme for Peace, Security and Development[2]), HN LFSO teams will adopt a hybrid approach that focuses on enhancing security, economic development, and governance.

To provide security, villages identified as good candidates for LFSO will host an HN LFSO team that organizes, trains, equips, and supervises a "village guard force" of locals who can man checkpoints

---

[1] Negative outcomes or counteracting effects are taken into consideration during the review of the Theory of Change below (see Figure 4.4).

[2] The Malian authorities intended for the Special Programme to curb insecurity, poverty, youth unemployment, hostage-taking, and all forms of trafficking. The approach was to construct 20 socioeconomic facilities for the populations of the Kidal and Gao regions, including health centers, schools, modern wells, and housing for officials (Touré, 2012).

and patrol in and around the village. The HN LFSO team, in turn, will function as a Quick Reaction Force to augment defenses when needed.

To enhance economic development, the HN LFSO team will focus on bringing in development funding, improving agricultural practices, improving roads, facilitating trade, and facilitating communication processes, such as intravillage dialogue and negotiations.

To enhance governance, the HN force will facilitate village stakeholder meetings, support inclusionary and transparent political processes, and link villagers to the HN government. For example, local mayors and elders will be encouraged to participate in decisionmaking and information collection. Additionally, village leaders will be encouraged to correspond with and visit higher-level government officials outside of their village.

In order to deconflict operations, the HN LFSO teams will coordinate with other actors in the operational environment: HN military units, foreign military trainers and Special Operations Forces, NGOs, and others.

The success of the U.S. contribution to the LFSO effort will be determined in large part by the ability of the HN to sustain LFSO over the long term and develop its own train-the-trainer capability so that it can in turn help neighboring countries implement stability operations at the local level.

Assessment Plan

As noted above, the HN has requested that the USEMB-based interagency team perform ongoing assessment of its LFSO efforts. For the purposes of this example, it is assumed that the USEMB has created a small assessment cell (e.g., one O-4/GS-13, one O-2/GS-11, one assessment SME contractor, and one intelligence noncommissioned officer, augmented by one HN representative who can add local context and insights) to cover that part of the mission. In the following, we describe the assessment plan for the notional LFSO effort outlined above to demonstrate how to apply the recommendations derived from this research.

The assessment plan contains the following steps:

1. Identify and address challenges specific to the scenario
2. Establish or review the Theory of Change
3. Determine metrics and how (and when) to collect them
4. Set up processes for data analysis, aggregation, and communication of results
5. Brief leadership and other stakeholders on the assessment plan (and adjust the plan if necessary).

Addressing and Mitigating Scenario-Specific Challenges

Assessment in this scenario is made more difficult by the complex mission design and the number of organizations involved in the effort—HN and U.S. government and civilian agencies, NGOs, local civilians, and other stakeholders. Like all LFSO, lines of effort are overlapping and influence each other. The regional context adds further complexity. The assessment team is therefore working closely with the mission planners to make sure the mission and its objectives are clearly defined. The assessment team is also engaging with SMEs early on in the process in order to develop a full understanding of the situation and the context in which the mission and the assessment will take place. This facilitates the generation of a comprehensive Theory of Change, which then is constantly reassessed and adapted in response to developments on the ground, taking into account positive as well as negative feedback loops.

In a complex mission, it is also challenging to strike the right balance between qualitative and quantitative measures. Collecting metrics comes with a cost, so the assessment team takes care to prioritize quality over quantity of metrics. However, the team knows that it must not only capture the three main lines of effort (security, development, and governance), it must also be able to track what is going on in the village and its wider surroundings beyond the LFSO effort. This is reflected in the selected metrics below.

The complexity of the mission not only makes it difficult to determine what to collect and assess, it also requires a clear distribution of labor regarding who collects and assesses. Metrics will be derived from a variety of sources, and some of the organizations involved are already performing their own assessments, which can be leveraged. However, the assessment team knows it must identify and mitigate the potential for biased assessments stemming from organizational self-interest.

Along with the distribution of labor comes a time-based aspect: maintaining the assessment effort over the years that LFSO will be taking place. The initial assessment plan therefore has provisions for continuing the effort after its originators have moved on, including a thorough documentation of the process and associated rationale, as well as periodic review and adaptation. The assessment team also prepares for future changes by collecting a certain amount of additional baseline data beyond what is required for the current plan. It also plans to educate each wave of LFSO teams about the assessment effort and what is expected of them in this context. This is done by including a brief presentation of the assessment during LFSO team pre-mission training and by asking the LFSO teams to discuss their contributions to the assessment process during handover from one to the next.

Finally, the remote and austere environment in which the LFSO takes place in this scenario makes data collection difficult and expensive. The assessment team therefore incorporates and utilizes existing data sources as much as possible and carefully determines which—if any—additional data must be collected.

Reviewing the Theory of Change

The assessment team expanded the initial Theory of Change for LFSO in this country (shown in Figure 4.3) by adding consideration of outside influencers that are not linked to the LFSO effort (the green boxes in Figure 4.4) and also included some potential spoilers based on an analysis of the mission and discussions with SMEs (the red text and red arrows in Figure 4.4). Understanding the importance of a consistent approach, the assessment team also shared these insights with the overall LFSO mission planning team.

Figure 4.4
Negative Outcomes/Effects and External Influences Added to the Basic Theory of Change

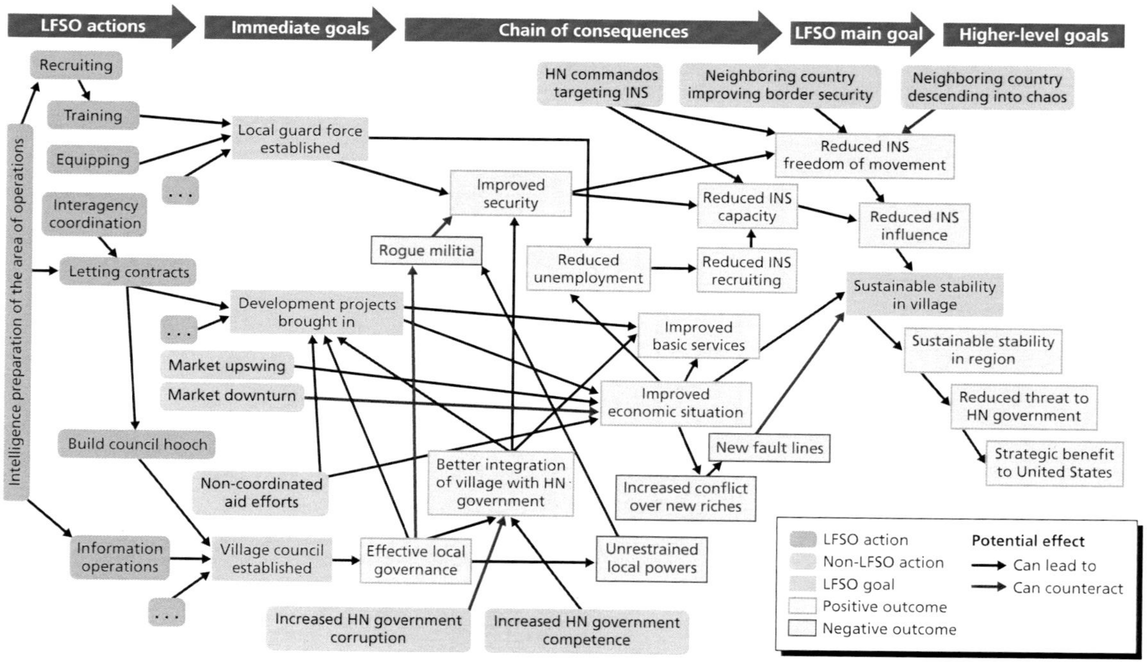

RAND RR387-4.4

Metrics and Their Collection

On the basis of the scenario and considerations outlined above and additional SME input, the assessment team created a list of metrics and associated documentation that will help them implement and manage the assessment effort. The metrics are organized by line of effort (security, development, governance, stability) in Tables 4.1 to 4.4 below. For each metric, the following information is provided:

- Metric type
  - Quantitative data: numbers (plain integer or floating point values), ratios (fractions), ordinal metrics (discrete, ranked categories, such as "high," medium," "low"), nominal metrics (discrete, nonranked categories, such as nationality)
  - Narrative data (pure text without quantitative information)
  - Mixed data (text containing quantitative information)
- Collection frequency: biweekly, monthly, or quarterly
- Who collects the metric
  - The assessment team itself, either through its own efforts or by pulling data from existing sources
  - The LFSO teams in the villages
  - Staff at the USEMB in the HN
  - The villager survey contractor/NGO
  - The joint HN/U.S. intelligence support team
- Rationale and additional remarks

Some metrics can be further broken down into submetrics or precursor metrics (e.g., the numerator and denominator of a ratio metric, as well as all mixed data). Figure 4.5 shows how these metrics are linked to the elements of the Theory of Change.

Planning for Data Analysis and Communication

The assessment team also designed a process to make sure the data analysis proceeds smoothly and the results have an impact on the LFSO effort. Each analysis cycle results in products for both "upstream" recipients, such as commanders and other decisionmakers, and "downstream" consumers, such as the HN LFSO teams and interagency

Table 4.1
Security-Related Metrics

No.	Metric	Rationale	Type	Collection Frequency	Collector	Remarks
1	Number of members of guard force that meet proficiency and equipping standards		Quanti-tative	Biweekly	LFSO team	
2	Number of guard force patrols per week	Contributes to feeling of security by villagers, deters INS, helps gather intelligence, and demonstrates guard force capability and capacity	Quanti-tative	Biweekly	LFSO team	
3	Number of INS-caused casualties among villagers		Quanti-tative	Biweekly	LFSO team	
4	Level of INS attacks in AO		Mixed	Biweekly	LFSO Team	Caveat: not a direct indicator of security situation
5	INS influence/propaganda/ intimidation efforts in AO		Mixed	Biweekly	LFSO team, intelligence team	
6	Number of calls to tip line from AO	Indicates level of intimidation, and pro-HN-government attitude of villagers	Quanti-tative	Biweekly	LFSO team	
7	Villager perception of security		Narrative	Quarterly (survey team), biweekly (LFSO team)	Survey team, LFSO team	

Table 4.1—Continued

No.	Metric	Rationale	Type	Collection Frequency	Collector	Remarks
8	Children moving around unaccompanied	Validates survey data about villager perceptions of security	Narrative	Biweekly	LFSO team	
9	Number of INS camps in AO		Mixed	Biweekly	LFSO team, intelligence team	
10	Level of guard force abuses		Mixed	Biweekly	LFSO team, survey team	
11	Number of military-aged males in AO	Gives perspective to guard force strength and (combined with unemployment rate) INS recruiting potential	Quanti-tative	Quarterly (survey team), biweekly (LFSO team)	LFSO team, survey team	
12	Whether village power brokers stay in village		Mixed	Biweekly	LFSO team	
13	Kill/capture operations by HN commandos in AO	Critical environmental factor	Mixed	Monthly	USEMB	
14	Security situation in neighboring countries	Critical environmental factor	Narrative	Monthly	USEMB	

Table 4.2
Development-Related Metrics

No.	Metric	Rationale	Type	Collection Frequency	Collector	Remarks
20	Cell coverage level	Level of basic services	Mixed	Biweekly	LFSO team, USEMB	
21	Cell coverage interruptions	INS-caused interruptions?	Quanti-tative	Biweekly	LFSO team, intelligence team	
22	Power-grid availability	Level of basic services	Quanti-tative	Biweekly	LFSO team	
23	Power-grid interruptions	INS-caused interruptions?	Quanti-tative	Biweekly	LFSO team	
24	Number of wells	Level of basic services	Quanti-tative	Biweekly	LFSO team	
25	New businesses	Indicator of investment activity	Mixed	Biweekly	LFSO team	
26	Number of market stalls	Indicator of investment activity	Quanti-tative	Biweekly	LFSO team	
27	Unemployment rate	Indicator of economic situation; also plays into security (INS recruiting potential)	Quanti-tative	Biweekly	LFSO team, USEMB	Compare with historic baseline and HN average
28	Outside aid activity	Tracking external influences	Mixed	Biweekly	LFSO team, USEMB	
29	LFSO-initiated aid contract volume	Direct activity indicator	Mixed	Biweekly	LFSO team	

Table 4.3
Governance-Related Metrics

No.	Metric	Rationale	Type	Collection Frequency	Collector	Remarks
40	Level of property conflicts	Rule-of-law indicator; also plays into investment climate	Mixed	Quarterly	Survey team	Narrative also needs to cover how conflicts are addressed/handled
41	Corruption issues		Narrative	Quarterly (survey team), biweekly (LFSO team)	LFSO team, survey team	
42	Number of members on village council		Mixed	Biweekly	LFSO team	
43	Number of effective meetings of village council per reporting period		Quanti-tative	Biweekly	LFSO team	
44	Villager participation in decisionmaking		Narrative	Quarterly (survey team), biweekly (LFSO team)	LFSO team, survey team	
45	Perceived legitimacy of village council		Narrative	Quarterly (survey team), biweekly (LFSO team)	LFSO team, survey team	
46	Necessary resources available to village council		Narrative	Biweekly	LFSO team	
47	Number of village council projects successfully completed		Quanti-tative	Biweekly	LFSO team	
48	Quality and type of interaction between village council and LFSO team		Mixed	Biweekly	LFSO team	

Table 4.3—Continued

No.	Metric	Rationale	Type	Collection Frequency	Collector	Remarks
49	Quality of interaction between village council and district/province government		Mixed	Quarterly	LFSO team	
50	Number of village cases in HN court system	Indicator of HN capability and strength of ties between HN government and village	Quanti-tative	Quarterly	LFSO team, USEMB	Interpretation depends on context
51	Percentage of convictions in court system		Quanti-tative	Quarterly	USEMB	
52	Number of cases handled by traditional village court	Indicator of strength of local community	Quanti-tative	Quarterly	LFSO team	Interpretation depends on context
53	Level of animosity toward "new rich"	Indicator of potentially developing fault lines	Narrative	Quarterly	Survey team	

Table 4.4
Stability-Related Metrics

No.	Metric	Rationale	Type	Collection Frequency	Collector	Remarks
60	Number of permanent residents in the AO/village	Influx or exodus of villagers is tied to the population's perceptions of overall stability	Quanti-tative	Biweekly	LFSO team	
61	Number of internally displaced persons in AO	Captures regional stability differential; increasing numbers of internally displaced persons may also lead to increasing instability	Quanti-tative	Biweekly	LFSO team	
62	External investment level	Indicator of external perceptions of long-term stability	Mixed	Biweekly	LFSO team	
63	Infant/child mortality	Indicator of health-related resources (also has security-related aspect)	Quanti-tative	Biweekly	LFSO team	
64	Villager perceptions of quality of life	Overall metric	Mixed	Quarterly	Survey team	
65	Villager perceptions of stability trend	Overall metric	Mixed	Quarterly	Survey team	
66	Villager concepts of "perfect world"	Serves to anchor context	Narrative	Annually	Survey team	Vignette format

Figure 4.5
Metrics Linked to Theory of Change

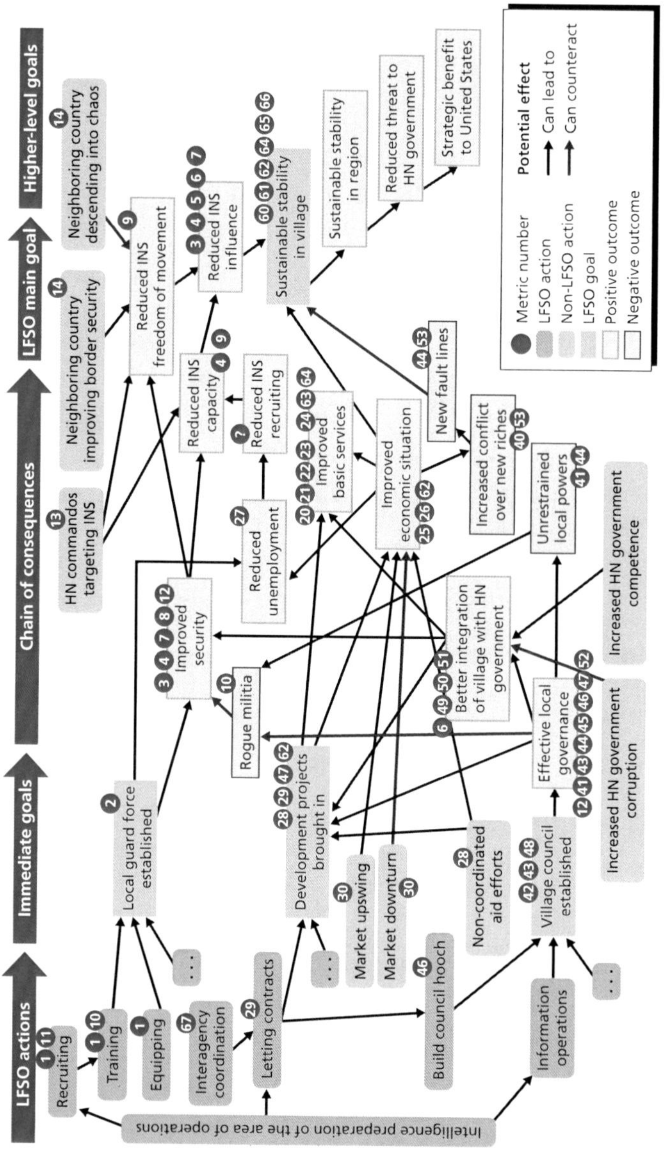

RAND *RR387-4.5*

action officers. For each product and recipient type, both "push" and "pull" distribution mechanisms are used.

In this case, commanders and other upstream recipients will receive brief monthly overview reports via e-mail (push delivery); these reports will reference more-detailed analysis products, as well as historical reports and data provided online (pull distribution).

The assessment-team lead will also participate in the weekly interagency planning meetings at the USEMB, where support for the HN's LFSO effort is coordinated, in order to track where the effort is going and to inject relevant results and insights from the assessment process.

HN LFSO teams will receive a customized report with results that are relevant to their AO on a biweekly basis, provided as hardcopy and on CDs, shipped with the resupply drop. The CDs will contain all other current and past products for reference, as well as the raw data, so that the LFSO team can dig deeper into the assessment if necessary. Because of the challenging communication situation, this type of physical delivery is chosen over methods that require a high-speed data connection. However, the assessment team also participates in the weekly status updates that the LFSO teams send via high-frequency radio link to provide time-critical assessment results orally and also answers any assessment-related questions the LFSO team may have.

Each assessment report contains contact information for the assessment team and a request for the recipient to provide feedback on the usefulness of the report so that the assessment team can adjust its efforts as needed. Selected products and data are also made available via a protected APAN website,[1] so that authorized users can pull assessment reports and data as required. In addition, the assessment team archives all reports and data on the knowledge-management systems of the USEMB and United States Army Africa.

---

[1] The All Partners Access Network (APAN) is a commercially secure, internet based, unclassified information-sharing portal with capabilities that enable virtual collaboration and coordination. It is designed to enable military forces to share information and to conduct collaborative efforts with nontraditional, nonmilitary partners.

CHAPTER FIVE

Conclusions

Findings

This study was undertaken to provide answers to the following research questions:

- What are the characteristic elements of LFSO?
- What are desired outcomes (ends) of such operations, and through what tools (means) can they be achieved?
- How can these outcomes and costs be measured (metrics), and how can these measurements be collected (methods)?
- How should the collected data be analyzed and the results communicated?

We developed a working definition of LFSO that addresses the current doctrinal gap: LFSO are the missions, tasks, and activities that build security, governance, and development by, with, and through the directly affected community in order to increase stability at the local level. We also identified several fundamental challenges and developed approaches to mitigating them.

Fundamental challenges to assessing LFSO include

1. The inherent complexity of LFSO missions
2. Limited assessment doctrine, training, and guidance
3. Competing visions of stability among stakeholders
4. The need to combine metrics and assessments across multiple areas and levels

5. Invalid or untested assumptions about causes and effects
6. Bias, conflicts of interest, and other external factors that create perverse incentives
7. Redundant reporting requirements and knowledge-management challenges
8. The difficulty of continuing the assessment process across deployment cycles
9. Failure to include HN perspectives.

Several themes run through these challenges. First, there is uncertainty or lack of consensus over how assessments should be conducted, what the mission objectives are, or how a command believes its activities will achieve its objectives. Second, incentives and goals among participants in the assessment process, ranging from subordinate commanders to local survey respondents, are not necessarily aligned with the goals of the assessment. Third, as a perhaps unavoidable artifact of the first two themes, integrating the perspectives and goals of all stakeholders is extraordinarily difficult.

That said, we have identified three principles that can help commanders and assessment teams address this daunting task:

- **Assessments should be commander-centric.** If the assessment process does not directly support the commander's decisionmaking, it should be revised.
- **Assessments should reflect a clear Theory of Change.** In order for the assessment team to adequately support the commander, it must understand not only the commander's objectives, but also the underlying Theory of Change—how and why the commander believes the tasks that have been laid out will result in the desired end state. A clearly articulated Theory of Change allows the assessment team to identify the appropriate inputs, outputs, and outcomes to measure, and also enables it to determine whether critical assumptions built into the concept of operations may, if proven faulty, require the commander to adjust the campaign plan.

- **Assessments should seek to triangulate the truth.** Assessment teams should fully exploit the data and methods available, leveraging the strengths of one source against the weaknesses of others to triangulate ground truth. The assessment team should exploit the healthy tension between quantitatively and qualitatively focused analysts and ensure that analyses are synchronized to support the commander's objectives, rather than simply being a science project that responds to reporting requirements of higher headquarters.

Finally, we have designed and demonstrated an assessment process that can serve as a template for teams tasked with assessing LFSO and similar operations. It includes the following steps:

1. Identify the challenges specific to the scenario
2. Establish the Theory of Change behind the planned operation to help document the expected results and describe how activities and tasks are linked to those results
3. Determine metrics and how and when to collect them
4. Set up processes for data analysis (including aggregation) and communication of results
5. Develop options for briefing leadership and stakeholders on the assessment plan.

Future Research

LFSO are a novel concept with historical precedent but without clear doctrinal definition or guidance. Additional work should be done to more sharply define LFSO (and to clarify what it is not), to identify environments and strategic objectives that might call for the use of LFSO, to identify the tactical conditions under which LFSO can be successful, and to suggest ways that LFSO might be conducted differently than in the past (e.g., HN-led with episodic U.S. engagement vs. U.S.-led and -staffed continuous effort).

Additional research should be conducted on how to tailor and adapt existing assessment tools to new environments. Many of the metrics identified (e.g., for Afghanistan) will be portable to other contingencies, but others will not be, and still others will require adaptation. Understanding how to treat these different metrics requires knowledge not only of the mission specifics but also of how cultural context affects the meaning of the metrics themselves. Additionally, real-world assessment requires triangulation (called *mixed methods* in the program-evaluation literature) to understand ground truth. Unfortunately, few formal tools exist for understanding how the synthesis of different sources of data can impact the level of confidence in the resulting inferences.

Finally, there are many remaining opportunities to increase the quality and accessibility of the assessment guidance and training that is available to commanders and assessment personnel. The Army should evaluate the best way to institutionalize assessment expertise. The development of joint assessment doctrine is one important step. If the Army chooses to make assessments an ORSA responsibility, ORSAs will require additional education and training to sensitize them to methods that fall well outside their current requirements (e.g., anthropology). While these broader institutional initiatives are under way, the Army should consider developing additional training products, from compact reference guides[1] to instructional videos, to help deploying analysts and commanders.

---

[1] For a useful example, see Center for Army Analysis, 2007.

References

162nd Infantry Brigade, "Operations Group & 162nd FSF-CA SFAT Training Concept," Fort Polk, La., December 2011.

Banko, Katherine, Peter Berggren, Aletta R. Eikelboom, Ivy V. Estabrooke, Trevor Howard, Roman Lau, Jenny Marklund, Joakim Marklund, A. J. van Vliet, and Andy Williams, *HFM–185 RTG: Processes for Assessing Outcomes of Multinational Missions*, North Atlantic Treaty Organization Research and Technology Organization, undated. As of June 3, 2013:
http://www.mors.org/UserFiles/file/2012%20-%20Meeting%20AMNO/HFM%20185%20final%20draft%2019%20Sept.pdf

Becker, David C., and Robert Grossman-Vermaas, "Metrics for the Haiti Stabilization Initiative," *Prism*, Vol. 2, No. 2, March 2011, pp. 145–158.

Bowers, David, "Which Tribe Should We Engage: A Tribal Engagement Assessment Methodology," *Small Wars Journal*, January 25, 2013. As of August 22, 2013:
http://smallwarsjournal.com/jrnl/art/which-tribe-should-we-engage-a-tribal-engagement-assessment-methodology

Bruck, Tilman, Patricia Justino, Philip Verwimp, and Alexandra Avdeenko, *Identifying Conflict and Violence in Micro-Level Surveys*, Bonn, Germany: Institute for the Study of Labor (IZA), 2010. As of June 8, 2013:
http://www.iza.org/en/webcontent/publications/papers/viewAbstract?dp_id=5067

Campbell, Jason, Michael E. O'Hanlon, and Jeremy Shapiro, "How to Measure the War," *Policy Review*, No. 157, October 1, 2009. As of May 15, 2013:
http://www.hoover.org/publications/policy-review/article/5490

Cancian, Matthew F., "Counterinsurgency as Cargo Cult," *Marine Corps Gazette*, January 2013, pp. 50–52.

Capshaw, N. Clark, and Jeffrey W. Bassichis, *AFRICOM's Campaign Assessment Process*, Military Operations Research Society, undated.

Center for Army Analysis, *Deployed Analyst Handbook*, Fort Belvoir, Va., 2007. As of August 22, 2013:
https://call2.army.mil/docs/doc3122/DAHB_Final_(8.5x11).pdf

Center for Army Lessons Learned, *Assessment and Measures of Effectiveness in Stability Ops*, Fort Leavenworth, Kans., May 2010. As of May 16, 2013:
http://usacac.army.mil/cac2/call/docs/10-41/10-41.pdf

Clancy, James, and Chuck Crossett, "Measuring Effectiveness in Irregular Warfare," *Parameters*, Summer 2007. As of June 20, 2013:
http://strategicstudiesinstitute.army.mil/pubs/parameters/articles/07summer/clancy.pdf

Connable, Ben, *Embracing the Fog of War: Assessment and Metrics in Counterinsurgency*, Santa Monica, Calif.: RAND Corporation, MG-1086-DOD, 2012. As of August 20, 2013:
http://www.rand.org/pubs/monographs/MG1086

Cordesman, Anthony H., Adam Mausner, and Jason Lemieux, *Afghan National Security Forces: What It Will Take to Implement the ISAF Strategy*, Washington, D.C.: Center for Strategic and International Studies (CSIS), November 2010.

Downes-Martin, Stephen, "Operations Assessment in Afghanistan Is Broken: What Is to Be Done?" *Naval War College Review*, Vol. 64, No. 4, Autumn 2011, pp. 103–125.

Eles, P. T., E. Vincent, B. Vasiliev, and K. M. Banko, *Opinion Polling in Support of the Canadian Mission in Kandahar: A Final Report for the Kandahar Province Opinion Polling Program, Including Program Overview, Lessons, and Recommendations*, Ottawa, Canada: Defence R&D Canada, Centre for Operational Research and Analysis, DRDC CORA TR 2012–160U, September 2012.

Groves, Robert M., Floyd J. Fowler, Jr., Mick P. Couper, James M. Lepkowski, Eleanor Singer, and Roger Tourangenau, *Survey Methodology*, Hoboken, N.J.: John Wiley & Sons, Inc., 2009.

Headquarters, Department of the Army, *Counterinsurgency*, Washington, D.C., FM 3-24, December 15, 2006.

———, *Stability Operations*, Washington, D.C., FM 3-07, October 2008.

———, *The Operations Process*, Washington, D.C., FM 5-0, March 26, 2010.

Kilcullen, David, "Measuring Progress in Afghanistan," Kabul, Afghanistan, December 2009. As of July 5, 2013:
http://literature-index.wikispaces.com/file/view/Kilcullen-COIN+Metrics.pdf

LaRivee, Dave, *Best Practices Guide for Conducting Assessments in Counterinsurgencies*, Washington, D.C.: U.S. Air Force Academy, December 2011.

Mausner, Adam, *Reforming ANSF Metrics: Improving the CUAT System*, Washington, D.C.: Center for Strategic and International Studies, August 2010. As of August 21, 2013:
http://csis.org/files/publication/100811_ANSF.CUAT.reform.pdf

Meharg, Sarah Jane, *Measuring Effectiveness in Complex Operations: What Is Good Enough*? Calgary, Canada: Canadian Defence & Foreign Affairs Institute, October 2009.

Military Operations Research Society (MORS), "Assessments of Multi-National Operations," PowerPoint presentation, MacDill Air Force Base, Tampa, Fla., November 5–8, 2012.

Odierno, Raymond T., James F. Amos, and William H. McRaven, "Strategic Land Power: Winning the Clash of Wills," United States Army, United States Marine Corps, and the United States Special Operations Command, 2013. As of March 16, 2014:
http://www.tradoc.army.mil/FrontPageContent/Docs/Strategic%20Landpower%20White%20Paper.pdf

Office of the Coordinator for Reconstruction and Stabilization, *The Interagency Conflict and Assessment Framework*, U.S. Department of State, Washington, D.C., c. 2008. As of August 22, 2013:
http://www.state.gov/documents/organization/187786.pdf

Office of the Special Inspector General for Afghanistan Reconstruction, *Actions Needed to Improve the Reliability of Afghan Security Force Assessments*, Arlington, Va.: OSIGAR, Audit 10-11, June 29, 2010. As of August 21, 2013:
http://www.sigar.mil/pdf/audits/2010-06-29audit-10-11.pdf

Office of the Under Secretary of Defense for Acquisition, Technology and Logistics, *Report of the Defense Science Board Task Force on Defense Intelligence Counterinsurgency (COIN) Intelligence, Surveillance, and Reconnaissance (ISR) Operations*, Washington, D.C., February 2011, pp. 27–28.

Organizational Research Services, "Theory of Change: A Practical Tool for Action, Results and Learning," 2004. As of March 19, 2014:
http://www.aecf.org/upload/publicationfiles/cc2977k440.pdf

PA Consulting Group, "Dynamic Planning for COIN in Afghanistan," briefing, 2009. As of July 14, 2013:
http://msnbcmedia.msn.com/i/MSNBC/Components/Photo/_new/Afghanistan_Dynamic_Planning.pdf

Paul, Christopher, Colin P. Clarke, Beth Grill, and Molly Dunigan, *Paths to Victory: Lessons from Modern Insurgencies*, Santa Monica, Calif.: RAND Corporation, RR-291/1-OSD, 2013. As of March 16, 2014:
http://www.rand.org/pubs/research_reports/RR291z1.html

Paul, Christopher, Harry J. Thie, Elaine Reardon, Deanna Weber Prine, and Laurence Smallman, *Implementing and Evaluating an Innovative Approach to Simulation Training Acquisitions*, Santa Monica, Calif.: RAND Corporation, MG-442-OSD, 2006. As of March 15, 2014:
http://www.rand.org/pubs/monographs/MG442.html

"Remarks by the President at the National Defense University," Fort McNair, Washington, D.C., May 23, 2013. As of December 20, 2013:
http://www.whitehouse.gov/the-press-office/2013/05/23/remarks-president-national-defense-university

Rossi, Peter H., Mark W. Lipsey, and Howard E. Freeman, *Evaluation: A Systematic Approach*, 7th ed., Thousand Oaks, Calif.: SAGE Publications, 2004.

Schroden, Jonathan J., "Measures for Security in a Counterinsurgency," *The Journal of Strategic Studies*, Vol. 32, No. 5, October 2009, pp. 715–744.

———, "Why Operations Assessments Fail," *Naval War College Review*, Vol. 64, No. 4, Autumn 2011, pp. 88–102.

Stevens, Stanley S., "On the Theory of Scales of Measurement," *Science*, Vol. 103, No. 2684, 1946, pp. 677–680.

Stewart, Chris, *Stakeholder Interviews Thematic Summary: Project Certain Echo*, Washington, D.C.: Gallup, January 24, 2013, Not available to the general public.

Survey Research Center, Institute for Social Research, "Guidelines for Best Practice in Cross-Cultural Surveys," Ann Arbor, Mich.: University of Michigan, 2011. As of March 15, 2014:
http://www.ccsg.isr.umich.edu/pdf/FullGuidelines1301.pdf

Touré, Boubacar Sidiki, "Sahel and West Africa," Organisation for Economic Co-operation and Development, June 2012.

Upshur, William P., Jonathan W. Roginski, and David J. Kilcullen, "Recognizing Systems in Afghanistan: Lessons Learned and New Approaches to Operational Assessments," *Prism*, Vol. 3, No. 3, June 2012, pp. 87–104.

U.S. Department of State, "U.S. Relations with Nigeria," Bureau of African Affairs Fact Sheet, Washington, D.C., August 28, 2013. As of December 20, 2013:
http://www.state.gov/r/pa/ei/bgn/2836.htm

U.S. Joint Chiefs of Staff, *Joint Operation Planning*, Washington, D.C., Joint Publication 5-0, August 11, 2011a.

______, *Joint Operations*, Washington, D.C., Joint Publication 3-0, August 11, 2011b.

______, *Commander's Handbook for Assessment Planning and Execution*, Version 1.0, Joint Staff, J-7, Joint and Coalition Warfighting, Suffolk, Va., September 9, 2011c. As of September 20, 2012:
http://www.dtic.mil/doctrine/doctrine/jwfc/assessment_hbk.pdf

Zukin, Cliff, "A Journalist's Guide to Survey Research and Election Polls," American Association for Public Opinion Research, 2012. As of May 18, 2013: http://www.aapor.org/AM/Template.cfm?Section=Journalist_s_Guide

Zyck, Steven A., *Measuring the Development Impact of Provincial Reconstruction Teams*, Civil Military Fusion Centre, June 2011.